Salmon, People, and Place

Salmon, People, and Place

A Biologist's Search for Salmon Recovery

Jim Lichatowich

Oregon State University Press
Corvallis

Library of Congress Cataloging-in-Publication Data

Lichatowich, Jim.

 Salmon, people, and place : a biologist's search for salmon recovery / Jim Lichatowich.

  pages cm

 Includes bibliographical references and index.

 ISBN 978-0-87071-724-6 (alk. paper) -- ISBN 978-0-87071-725-3 (e-book)

1. Pacific salmon fisheries--Northwest, Pacific--Management. 2. Pacific salmon --Conservation--Northwest, Pacific. I. Title.

 SH348.L528 2013

 333.95'656--dc23

2013013094

Oregon State University Press

121 The Valley Library

Corvallis OR 97331-4501

541-737-3166 • fax 541-737-3170

www.osupress.oregonstate.edu

Dedication

For Paulette, Jim, Sue, Tim, and especially for Ella and her generation.

In Memoriam

Daniel Callaghan (1931-2006) and Steve Bukeida (1946-2006)
Two strong advocates for wild salmon and steelhead.

No doubt one of the things that the history books of the future will record is the present awakening of men to the value of the other living things that share his world. The vast supply of timber, fur bearing animals, fish and game, is no longer considered to be inexhaustable. It was a tragic thing to sit idly by while certain creatures were completely exterminated; it is almost as tragic today to witness the destruction of the haunts of wild life. Many North American species have been pressed back to the frontiers or to preserves where they are little more than museum pieces or rare curiosities. The Atlantic salmon is now gone from most of the New England rivers while across the continent the Pacific salmon is waging a losing fight. It seemed to be undoubtedly true that the public ownership of wild life had been a failure. But now there is a different spirit abroad and it remains for public management to prove, wherever wild life still exists, that it can master the situation.

—C. McC. Mottley, Foreword in Roderick Haig-Brown's
The Western Angler. 1947

The twenty-first century needs a new story about all the fish in the sea, a story that emphasizes their fragility.

—Carmel Finley, *All the Fish in the Sea: Maximum Sustained Yield and the Failure of Fisheries Management.* 2011

Table of Contents

Acknowledgements

This book would not have been written without the encouragement and help of my wife and best friend, Paulette. The encouragement from the rest of my family—Jim Jr., Sue, Ella, and Tim—was also a great help. Much of the motivation to write the book came from them.

A very special thanks to Jim Hall, whose detailed review of the manuscript and editorial suggestions greatly improved the book. I would like to thank Mary Braun for her encouragement and assistance in publication of this book. Jo Alexander's editing made substantial improvements in the manuscript. Everyone at OSU Press has been very helpful.

I was very fortunate that early in my career I was adopted by four mentors: Harry Wagner, Jack McIntyre, Charles Warren, and Homer Campbell. Any success I may have had in my career I owe to their guidance.

Special thanks to Willa Nehlsen and Jack Williams. Their invitation to work on the Crossroads paper set me on a path that led to my two books.

I appreciate Jim and Sharon van Loan's advice, support, and retreat times at the Steamboat Inn on the North Umpqua River.

I was also very fortunate to have spent long hours talking with many salmon experts. The ideas in this book had their genesis in conversations with Dan Bottom, Dave Buchanan, Steve Johnson, Dave Hankin, Jay Nicholas, Barry McPherson, Stan Gregory, Bill Pearcy, Ron Hirshi, Eric Laudenslager, Dave Philips, Lars Mobrand, Rick Williams, Jack Stanford, Bill Liss, John Esler, Carmel Finley, Frank Moore, Bill Bakke, Pete Soverel, Guido Rahr, Spencer Beebe, and Kurt Beardslee of Wild Fish Conservancy. Mr. Ron Allen of the Jamestown S'Klallam Tribe allowed me a glimpse at his native culture and Matt Murphy of Sherkin Island, Ireland, showed me how a family can grow and prosper while living in harmony with the land.

Thank you, also, to all the volunteers and watershed councils for their many hours of work. You are a treasure.

Introduction

This is the essence of tragedy, to have meant well and made woe.

—Robinson Jeffers[1]

Experience has a bitter taste in fisheries management.

—Callum Roberts[2]

Rain—steady, heavy rain—so much rain that air and water merge into a single, wet grayness. I'm listening to the steady drumming of raindrops on the hood of my jacket while I look over the clear-cut. The ground is covered with a tangle of limbs and the trunks of non-merchantable trees. Near the center of this field of logging debris is a small stream and it's that stream that brought me here today, to see if the loggers left the required number of trees along the stream bank. One glance at the clear-cut tells me what I need to know. All the riparian trees are gone.[3] About a dozen small trees scattered across the clear-cut did manage to escape the chain saw. The largest are no more than three to five inches in diameter.

I head toward the stream even though it's clear that the rules intended to protect it were ignored. Crossing the clear-cut is like running an obstacle course. My feet rarely touch the ground as I slip and stumble on the rain-slicked branches and logs. It is a small miracle that I fall only once before reaching the stream.

It's a small stream. There are places where I can stand with a foot on each bank, but it is home to several cutthroat trout, and now, with the riparian trees gone, their home is in jeopardy. This is a clear violation of the rules. I'll complain—and so will others—even though I already know the response: "The lumber milled from those trees has more economic value than the few little trout that might be affected." That the loggers broke the rules doesn't carry much weight out here on the west end of the Olympic Peninsula. Even if the forest managers could be convinced that what happened here is a mistake, even if they could be persuaded to use a different yardstick to measure the value of the forest, the stream, and the trout, nothing can remedy the problem that this violation creates for the fish. It will take at least fifty to eighty years to replace the trees that once protected this stream.

I linger on the stream bank for several minutes thinking about the cutthroat's fate, until a trickle of cold water finds its way past my rain jacket and runs down my back. The cold water abruptly brings me back to the present and it reminds me of the hard slog back to the truck. When I turn to leave, I see something that freezes me in place. Across the creek, about forty or fifty yards away, is one of the few surviving trees—a western redcedar about two or three inches in diameter and probably fifteen or sixteen feet high. The spindly cedar is bent over so that its top nearly touches the ground. What grabs my attention is a golden eagle perched on the highest point of the arch formed by the bowed cedar. We stare at each other through the rain and across the debris that was once a forest. I've seen many eagles, but never on such an unlikely perch. Maybe the limbs and trunks strewn across this clear-cut are the remains of the eagle's home.

⤚⤳

It has been a couple of decades since the eagle and I stared at each other through the rain. The image is still fresh and returns to my mind's eye from time to time. It's one of those events that become a haunting memory, like a burning question whose answer always lies just beyond the next experience.

While driving home that day, I continued to think about the eagle. I recalled a passage from Barry Lopez's book *Of Wolves and Men,* in which he describes an intense stare between predator and prey; a silent exchange of information that either triggers an attack or breaks it off. Lopez called it "the conversation of death" and pointed out that this silent communication can only occur between wild animals. Domesticated animals have had it bred out of them.[4]

Was the strange encounter with the eagle a different kind of conversation, a conversation of blame, reproach, or rebuke? For several days, I couldn't shake the feeling that it was more than a chance meeting of two species in the remains of a shattered forest. The distance between us was only forty or fifty yards, but the distance between our species was a chasm of insurmountable distance, a chasm made wider by my culture's belief that humans are not part of nature. Unlike the Native Americans, who freely talked to nature, my culture has built impenetrable barriers to such communication. But what if those barriers didn't exist? What story would the eagle tell? Why was she there on that bizarre perch showing no obvious fear of being so close to a human?

Not long after that day, I began the intensive study of the history of salmon management, which led to my book, *Salmon Without Rivers*. As I ventured deeper into the history of natural resource management, the eagle on that bowed cedar tree would occasionally emerge from some back corner of my brain. Over time, the image of the eagle on its pathetic perch and the spindly cedar bent under the eagle's weight became for me a symbol for the state of the region's natural resources. Those resources constitute an important part of our human habitat, our home, and it is bending under the weight of blind allegiance to destructive myths. The plight of the wild Pacific salmon is just one outcome of the poor stewardship of the Pacific Northwest's natural resources.

In their pristine abundance, the salmon were one of the natural wonders of North America. They were on par with the passenger pigeon, the great white pine forests, the plains buffalo, the redwood forests, and the California sardine. In the Pacific Northwest wild salmon were considered a keystone species—one that plays a critical role in maintaining the structure of an ecological community— because their large annual runs transferred nutrients from the ocean to the inland aquatic and terrestrial ecosystems of the Pacific Northwest. Those nutrients were important to the Native Americans as well as the eagles, bears, and many other species. The wild salmon were the silver threads that held together the complex tapestry of ecosystems that comprise the Pacific Northwest. Salmon were of such importance that they acquired the status of a regional icon.

Today, with the benefit of our stewardship, wild salmon have been protected and enhanced right out of their ancient habitats and into fish factories (hatcheries). The fish factory has largely replaced the ecological relationships that once sustained the wild salmon. As a result, the flow of nutrients from the ocean to the headwaters has, in many years, been reversed. The total weight of juvenile salmon released from fish factories is greater than the total weight of returning adult salmon.[5] Silted gravel, warm and polluted water, massive irrigation withdrawals, dams and other barriers, and overharvest are also a legacy of our stewardship.

For most of my seventy-two years I dealt with, studied, and described the landscapes and resources handed down to us from previous generations. But now that I am on the other side of the hill, I worry more and more about what I have added to or subtracted from that inheritance. I am thinking more and more about those who will live with the consequences of my generation's stewardship. This has caused me to worry about the landscape and resources that my granddaughter, Ella, and her generation will inherit. Such worry is the motivation for this book.

The people of the Pacific Northwest still place a high value on Pacific salmon. Even the yardstick used by the market economy—the amount of money spent— clearly gives high value to the salmon. The region is now investing a billion dollars a year in salmon recovery.[6] However, this level of investment is yielding few positive results. We will continue to spend massive amounts of money and get little in return until the obstacles to recovery are identified and removed. I hope this book contributes to that task.

The story told here is not a history, although some history enters the narrative. It is not a scientific monograph, although I do include information from scientific publications. It is not a memoir, although I do discuss several events that occurred during my forty years as a salmon biologist. Using history, science, and memoir I describe what the salmon have taught me about their problem, and I project that understanding into the steps we need to take to create a sustainable relationship with the region's icon. The story is divided into two parts, each with a short introduction: Part I has four chapters that describe the salmon's problem, what is preventing salmon recovery, and why the billions of dollars spent on wild salmon restoration programs have not been effective. Part II uses two chapters to describe what needs to be done to remove the impediments to salmon recovery. Each chapter is followed by a short essay. I call these essays side channels. In a watershed, side channels diverge from the mainstem of the river, but they contribute to the overall productivity of the watershed. In this book, the side channels are short essays that diverge from the main theme of the preceding chapter, but add important insight to the book's overall story.

Solving the salmon's problem requires a deep and critical examination of some of our fundamental assumptions about nature. Some will welcome such an examination. I hope they find what is presented here useful, but more importantly

I hope they expand and improve on it. Others, who have become comfortable with the status quo, will resist such an examination. They are content to continue to pursue policies that have already created more than a century of failure. They remain in the comfort of the status quo by spinning a tale of success. For more than a century the conventional wisdom was too comfortable to begin the difficult search for an alternative because, as the economist John Kenneth Galbraith explained: "It's easy to see why the conventional wisdom resists so stoutly such change. It is a far, far better thing to have a firm anchor in nonsense than to put out on the troubled seas of thought."[7]

Part I—Icebergs, Myths, and Stories

We are accustomed to think of myths as the opposite of science. But in fact they are a central part of it: the part that decides its significance in our lives.
—Mary Midgley[1]

Suppose a fact is inconsistent with the frames and metaphors in your brain that define common sense. Then the frame or metaphor will stay, and the fact will be ignored.
—George Lakoff[2]

In his book *Arctic Oil*, John Livingston uses an iceberg as a metaphor for environmental problems, because environmental problems, like the iceberg, can be divided into a small, visible part and a larger, hidden mass.[3] Livingston calls the exposed part of the environmental iceberg the issues; they are the highly visible effects of human activities. For Pacific salmon, the visible issues include dams, poor logging practices, over-grazed riparian vegetation, dewatered and polluted rivers, poor hatchery practices, and overharvest. The issues have been described and analyzed in scientific reports and in the popular media, especially since the widespread listing of Pacific salmon under the federal Endangered Species Act (ESA).

The issues are an important part of the salmon's problem, but, like the iceberg, the problem also has a large hidden component that is rarely the subject of scientific inquiry and never described in the popular media.[4] The submerged mass of the environmental iceberg hides from easy view the myths, assumptions, and beliefs that govern our behavior toward the natural world, and is the source of the decisions that create the visible issues. Various authors have given different names to the myths and assumptions that underlie resource management; they include story, conceptual foundation, frame, paradigm, and worldview. Although different names are used, they all refer to John Livingston's hidden part of the environmental iceberg. In this book, I primarily use story and conceptual foundation.

Myths? Assumptions? Do they really play a critical role in salmon man-agement? After all, we live in a modern society—dependent on advanced technology—that believes science is the source of truth about the world. Myths are commonly thought of as relics from primitive cultures that are naïve or passé—important only to the uninformed or gullible.[5] The Native American story of the Salmon King and his subjects who lived in five houses in the ocean fits this common understanding of myth.[6] However, myths are also systems of metaphors, symbols, assumptions, and beliefs that are used to interpret the stream of often confusing or contradictory signals emanating from the world around us. The machine is one of those symbols. It is a powerful metaphor used to describe the movement of celestial bodies, the functioning of ecosystems, and even the human body. Thinking and talking about ecosystems as though they were machines is a modern myth. It is a generally held assumption that our myths are different because they are derived from enlightenment values and the methods of science.[7] They avoid the stigma of the naïve story—or so we believe.

Myths selectively filter information. Philosopher Mary Midgley says, "The way in which we imagine the world [our myths] determines what we think is important in it, what we select for our attention among the welter of facts that constantly flood upon us. Only after we have made that selection can we start to form our official, literal, thoughts and descriptions."[8] An example from salmon management illustrates the selective filter of modern myth. For the past forty years fishery scientists have published report after report showing the harmful effects of hatchery programs on wild salmon. Yet even today, I still hear advocates for fish hatcheries say there is no evidence of the harmful effects of fish factories on wild salmon. When confronted with the overwhelming evidence that hatcheries have harmed wild salmon, they say the benefits outweigh the costs, although a real weighing of the costs and benefits has yet to be done. Outdated and erroneous myths retain their power to influence only as long as they remain hidden and protected from contrary facts and critical evaluation.

Salmon management agencies do employ the latest technology: genetic analysis, geographic information systems (GIS), the latest design in fish traps, and computerized storage and retrieval of all kinds of information. Those things are tools and they help salmon managers work toward a purpose, but they don't determine that purpose. Because the conceptual foundation filters and interprets information, it plays a powerful albeit unrecognized role in determining the purpose of management. This power to filter information can and does inhibit the transfer of new science into management and recovery programs when that new science is at odds with the conceptual foundation.[9] It distorts the perception of the salmon's problem and funnels recovery funds into ineffective projects.[10] As a consequence, the conceptual foundation has been perpetuating rather than solving the salmon's problem. Its underlying myths, beliefs, and assumptions about nature give salmon management a weak foundation upon which we have built a house of cards. To appreciate the importance of conceptual foundations, think of them as similar to the picture on a box containing a jigsaw puzzle. Each piece of the puzzle is a bit of information, but that information can only be interpreted by referring back to the picture on the box. Now imagine a puzzle that has the wrong picture on the box. For example, the picture on the box is a bouquet of flowers, but the pieces of the puzzle, when assembled, portray a sailboat on a stormy sea. The information on each piece of the puzzle, when compared to the picture, will either be misinterpreted or it may be judged irrelevant and discarded. There is little chance the puzzle will be completed. Salmon management biologists must interpret a steady stream of information from research and monitoring programs and a host of journal articles and reports. Those bits of information are the pieces of the salmon management puzzle. If the myths and assumptions about nature that make up the conceptual foundation give a false picture of the salmon's ecosystem and its processes, a lot of relevant pieces to the salmon management and recovery puzzle will be misinterpreted or ignored.[11] Some, for example, will ignore the evidence that hatcheries have failed to achieve their goals and that they have contributed to the decline of wild salmon.

The current approach to salmon recovery reminds me of an often-repeated comedy skit. A machine hidden behind a wall produces widgets and sends them into another room on a conveyor belt. A worker standing at the end of the conveyor belt is assigned to do something with the widgets, but becomes hopelessly tangled in a chaotic mess as the machine behind the wall speeds up and produces widgets faster than the worker can deal with them. The worker stays at the end of the conveyor belt flailing away, too occupied with the growing mess to go into the other room and correct the problem at its source. Our assumptions, myths, and beliefs are creating a continuous stream of environmental problems. State, federal, and tribal biologists, as well as private environmental groups work at the end of the conveyor belt flailing away at those problems. They are too occupied to attack the problems at their source.

Biologists attacking the visible issues are doing very important work. It is also critical that they step back from time to time and examine salmon management's conceptual underpinnings, which need continuous revision consistent with new scientific understanding of the salmon's natural production systems. We cannot hope to recover salmon if we continue to focus all our attention on the visible problems at the end of the conveyor belt while ignoring the source of those problems.

The four chapters that make up Part I look deeper into the hidden parts of the salmon's problem. Chapter 1 examines our sense of place and the powerful cultural filters that determine what we see and consequently how we behave towards this place we call home. Chapter 2 examines the assumptions, beliefs, and myths that underlie salmon management and recovery programs and how the current conceptual foundation has, for over a century, put salmon on the path to depletion and extinction. Chapter 3 examines the structure of institutions charged directly or indirectly with salmon conservation and how that structure reinforces the outdated and erroneous conceptual foundation. Chapter 4 recaps the first three chapters. It extends the approach currently followed by salmon management into the future and gives a glimpse of the salmon's fate if we continue on the current path.

A final note on conceptual foundations

A few months ago, I spent several evenings visiting an old friend—Aldo Leopold's *A Sand County Almanac*.[12] Every time I reread that wonderful book, I gain an enhanced understanding of conservation. I must have been thinking about conceptual foundations as I read, because for the first time I realized that in the last two sections of the book, "The Upshot" and "A Taste for Country," Aldo Leopold was describing his conceptual foundation, his beliefs and assumptions about nature. The rest of the book is a beautiful narrative that describes Leopold's natural world as he saw it through the lens of his conceptual system. Next time you read *A Sand County Almanac,* start at the back of the book and study the conceptual system described there, then work your way to the front. Keep in mind as you read the seasonal essays that you are seeing the world as Leopold saw it—filtered through the lens of his conceptual foundation. May we all have such a lens to look through some day.

Chapter 1—Winter Wrens and Jumbo Jets

Place ... refers to not just geographical locations ... but rather to an evolving, interactive relationship between an ecological system and the people who ... use it as their habitat.

—Bryan Norton[1]

More even than earthlings, we are placelings. Place is there to be seen only if we have the vision to behold it.

—Edward Casey[2]

The logging road winds through a cool, green tunnel of Douglas-fir, western redcedar, and western hemlock. The air is heavy with the rich, earthy smell of the rainforest. I can't resist the urge to step out of the pickup and fill my lungs with the intoxicating freshness. Even though most of the ancient forests on the Olympic Peninsula have been logged, it's still possible to find isolated patches of older trees, not the very old giants that were once abundant here, but trees that are maybe one hundred to two hundred years old. The firs and cedars in this patch of forest fall into this category and, even though in tree time they are just teenagers, they are big and impressive.

Usually, when I find a stand of these older trees, I look for a big Douglas-fir with a thick cushion of moss at its base. I sit down, lean back against the trunk, and listen to tree talk, the creaks, sighs, and groans of the big trees. I imagine that I'm hearing very old stories of strong winds, lightning, fires, long winters of soft, continuous rain, and dry summers. What tree wisdom do these giants accumulate during their long lives and what might we learn from them?

Tree talk is not as far-fetched as it seems because plants did evolve ways to communicate. For example, plants attacked by herbivorous insects emit phenolic compounds that attract predatory insects. The predators then eat the invading herbivores. This is an example of "plants talking to their body guards,"[3] one among the thousands of relationships that support the members of an ecosystem in their common goal of survival. But I cannot linger on these thoughts today. Living trees are not what brought me here. I am here because the owners of this land are converting the forest into two by fours and the clear-cut I am looking for is just up the road.

Clear-cuts are the prized tool of industrial foresters throughout the Pacific Northwest. Clear-cut logging and hatchery-dependent salmon runs have a common source. They come from the same underlying assumption that humans can enhance the production of commodities from ecosystems, if the natural systems are simplified and brought under human control.

This clear-cut probably started in an office in Forks or Port Angeles with a forester drawing innocuous-looking lines on a map. The lines didn't change the map much, but when my truck crosses one of those lines it's obvious and it's not innocuous. The forest abruptly ends; the cool air and subdued light immediately change to bright sun and a noticeably elevated temperature. A tangle of branches and other non-merchantable tree parts cover the ground on both sides of the road, all of it baking in the sun. I stop again, but it's not the natural slope and undulations of the mountainside that catch and hold my attention, it's the straight line that cuts across the landscape. On one side of the line is a mature forest and, on the other side, small rhododendrons appear to be the tallest living plants.

The old Ford pickup bucks and lurches over deep ruts and logging debris while I hang on to the steering wheel trying to guide two tons of bouncing steel toward the center of the ridge. Momentum carries the truck a few more yards before it stops in the middle of a landing—the place where logs were stacked before being loaded on trucks and taken off the mountain.

For three years, from 1988 to 1991, I spent most of my days in the backcountry of the Olympic Mountains away from the well-traveled roads, working as a habitat biologist for the Jamestown S'Klallam Tribe. My job was to check on logging operations across the northern Olympic Peninsula as part of Washington State's Timber, Fish, and Wildlife Program. Tribal and state biologists reviewed proposed timber harvests and recommended measures to protect stream habitats. They then returned to the site after it was logged to evaluate the results. Too often the recommendations had been ignored. The privilege of spending my days in the rugged and enchanting landscape of the Olympics came at the cost of almost daily encounters with the ecological consequences of a single-minded pursuit of profit. I'm sure that clear-cut logging is justified in the ledgers of accountants. I've never seen those ledgers, but I'm also sure they do not include a column for ecological costs. I have seen too many of those costs ignored or given minimal attention. After one particularly discouraging day another biologist summed up our inability to adequately protect salmon habitat by saying that salmon biologists do not prevent problems; we are merely scribes recording their occurrence and documenting their consequences.

I am on this ridge today to observe and record. A cedar stump about four feet high and three or four feet in diameter stands a few feet from where I parked the pickup. The tree was cut the first time this hillside was logged, probably early in the twentieth century. The stump is defiantly resisting its ultimate fate, although its edges are beginning to crumble. Its roots still penetrate and hold it to the earth from which it once drew its nourishment. The old stump's roots also penetrate time and hold onto a story. It's a story of ancient forests in a coevolving landscape. Once the glaciers retreated and freed the land from their icy grip, trees pushed their roots into the barren soil and took hold of it. Once established, the forest stabilized the mountains and riverbanks. Rivers protected by old-growth Douglas-fir and western redcedar welcomed the salmon. Fish and trees flourished. In time, forests of massive trees and massive runs of salmon created a unique place on earth. For thousands of years, the abundant runs of salmon were the basis for the Native American economies in the coastal and inland areas of the Pacific Northwest.

As I consider the stump, I am reminded that cedar was also an important part of the Native American economy, almost as important as the salmon. Hilary Stewart called it the tree of life in recognition of its importance. Cedar found its way into every part of the Indian culture and economy. The people built houses and canoes, carved totem poles and ceremonial masks, made paddles, boxes, bowls, fishhooks, cradles, fishing floats, salmon weirs, arrow shafts, baskets, and clothing out of cedar.[4] The Pacific Northwest was rich in natural resources and prior to the arrival of Euro-Americans: "the richest people in North America were Indians of the Northwest coast."[5] However, natural resources such as the cedar and salmon were more than just trees and fish, more than just commodities to the Native Americans.

Their stories and myths described the living world as a community of beings that included humans, cedar trees, salmon, clams, rivers, and eagles; each member of the community had equal status. In Native mythology animals talked to humans and they sometimes took human form. The coequal relationship between humans and the rest of nature had a powerful effect on behavior because "[p]eople do not exploit a nature that speaks to them."[6] Because they lived in a community whose members included all of nature and in which each member had equal status, the Native Americans had no special right to exploit salmon for food or profit. The salmon and other members of the community willingly offered themselves to humans as a gift, but the gifts would be forthcoming only if they were treated with respect.[7] The exchange of gifts among members of the community created a web of relationships that persisted for at least four thousand years. For Native Americans, the overarching metaphor for natural resources was the gift.

A little over a hundred and fifty years ago, men from a different culture began flooding into the Pacific Northwest. The newcomers had a very different vision of the natural world. Their arrival in the Pacific Northwest ushered in a rapid and wholesale change in the relationship between humans and nature. Euro-Americans believed humans were separate from nature and that they were ordained to exploit it for their personal comfort and profit. Technology such as the steel axe and saw, steam donkeys and locomotives, ships and canneries reinforced the Euro-Americans' separation from nature and strengthened their ability to exploit it. They used their technology to cut down the giant cedars and firs, to strip the rivers and mountains of gold and other minerals, and to capture and can salmon. Waste and overexploitation were acceptable, if they made money. Historian William Robbins described the change this way: "The acquisitive and aggressive newcomers brought with them a very different set of values, precepts that attached special significance to the manipulation and transformation of the physical world for personal gain."[8] Allerdale Grainger used the rough language of a turn-of-the-century logger to describe the new approach to natural resources: "As for him [a logger], he was going to butcher the woods as he pleased. It paid!"[9] For the Euro-Americans, the overarching metaphor for natural resources was money.

When the metaphor for natural resources changed from a gift to money, the rules of behavior also changed. It was every man for himself in the scramble to convert the forest, minerals, grass, salmon, and water to cash. The single-minded quest for profit was viewed as highly rational behavior, whereas goals not defined by monetary profit were considered irrational.[10] Consequently, the early salmon managers sympathized with the commercial fishermen's need for profit and they allowed overharvest. During the span of my own career, biologists trying to maintain the economic value of the fishery "scientifically" justified overharvest. The same attitude that ranked profit over stewardship considered it rational to exchange natural salmon habitat for fish factories.[11]

The shift of metaphor for natural resources from gift to money made the overexploitation of natural resources to maximize profit perfectly rational. A man in London can buy forestland in Oregon or Washington, land that could produce a sustained yield for generations, and then convert it all to cash in just a few years. Cutting all the trees in a short period of time is a wise market transaction to him. He is not touched by the negative consequences of his decisions: the destruction of local places, such as the little towns and communities supported by logging or the destruction of salmon-bearing streams that support fishing communities. Yes, this stump has a lot of stories.

Before the loggers cut down the cedar, they chopped a pair of deep notches into the tree about two feet above the ground. I can see inside the notches where the axe took out rough bites of the wood. The day that the cedar was killed, men drove long planks called springboards into the notches. They stood on the springboards as they pushed and pulled their saw through the tree, ending its centuries-long struggle to survive.

A winter wren flies from the forest at the edge of the clear-cut and lands inside one of the springboard notches. Winter wrens are usually the first forest residents to check me out when I visit places like this. The little ball of chocolate-brown feathers standing in the cave-like notch reminded me of its scientific name *troglodytes*, the cave dweller. It lets loose with its long, shrill, and complex song that sounds like it's scolding me for the devastation of its home. I say out loud, "It wasn't me." I suppose we all look the same to the winter wren, those who cut the trees and those who come later to record the consequences.

It's a beautiful day. The early morning sky is pale blue, clear and clean after recent heavy rains. The Strait of Juan de Fuca is a blue jewel set between the green hills of Vancouver Island and the Olympic Peninsula. A set of contrails arch across the sky—tracks of a jumbo jet cruising west. From my perspective the winter wren standing in a springboard notch a few feet away and the jumbo jet at thirty thousand feet appear the same size. I draw some satisfaction from the perspective that sets a few ounces of tissue and feathers equal to tons of metal. Both can fly, but that is their only similarity. One of the biggest differences between them is their relationships to place.

The wren lives in the understory and on the floor of older forests, rarely moving into the upper canopy. It is intimately connected to its habitat—the rotting log where it dines on insects, the hollow branch where it hides its nest, and the cedar stump where it sings. If it were possible to look through the wren's eyes at its habitat map, we would see a web of pathways connecting it to its home place. Those pathways and the wren's intimate knowledge of them are key to its survival. A disturbance such as a small lightning-caused fire or a mature tree blown over results in the need for an immediate adjustment in the wren's habitat map. The wren can adjust as long as the disturbance is not outside the range of its species' evolutionary experience. How will the wren adjust to a disturbance as large and complete as this clear-cut? It's not a matter of adjusting a few pathways; the whole habitat map has been erased.

For the distant owner of this land, the rules of the global economy severed the feedback loops that might have connected him to this mountain and its forest. For him, the only consequence of this clear-cut will be a small adjustment in the numbers on his accounting sheet. His connection to this land is as weak as the connection those people in the jumbo jet feel looking down from thirty thousand feet.

The jumbo jet is a poster child for our culture's dependence on highly advanced technology, a technology that increasingly separates us and insulates us from nature. The people on the plane depend on technology's power to keep them alive; they would die but for the artificial environment maintained by the plane's mechanical systems. The jumbo jet represents an abstract, artificial, and detached perception of habitat or place. Unlike the habitat maps of the winter wren, the abstract maps used by the pilots do not correspond to or recognize the real attributes of the places passing beneath the plane. The pilot's maps are useful navigational tools, but it would be a big mistake to confuse them with real places.[12] The jet carries its passengers from Seattle to Hawaii only because we have covered the world with abstract lines of latitude and longitude. The pilot knows exactly what spot on earth the plane is crossing at any time in relation to those lines. The jumbo jet is more than an isolated piece of machinery. It is a symbol of our culture's fondness for using technology to control a nature that has been simplified.

Technology not only separates humans from nature, it destroys the survival pathways of the residents of whole ecosystems. Once those ecosystems are degraded or destroyed, technology is then used to engineer substitute habitats for the plants or animals considered marketable commodities. For example, dams converted the Columbia River into a series of lakes and eliminated or greatly altered the salmon's historical pathways through its riverine habitat. Fish factories then replaced the need for those pathways, or so we thought, by taking the fish out of the river and rearing them under feedlot conditions, completely under human control. Our attempt to replace habitat with engineered substitutes overlooks the importance of relationships—relationships between animals and their habitat and relationships among the species of plants and animals in the ecosystem.

Salmon biologists talk about the importance of ecological relationships, but in practice they rarely incorporate them into their management and recovery programs. For example, the important relationship between salmon as carriers of nutrients from the ocean and the freshwater food webs that make use of those nutrients is rarely considered when salmon harvest managers determine how many fish will be caught and how many will be allowed to escape to the spawning grounds. The ecological relationships that sustain the salmon are ignored because they are not readily visible. They are not as easily counted as are salmon crossing a dam. Because relationships are not directly visible, they must be constructed, like a puzzle, from informed observations and data collected in well-designed sampling programs, all of which presupposes a conceptual foundation, a picture of the salmon-sustaining ecosystem that admits the existence and importance of relationships. An inadequate conceptual foundation renders important ecological relationships invisible.

Our lack of attention to ecological relationships is a major contributor to the precarious state of the salmon today. Biologist Gary Nabhan says that a species does not go extinct because someone shoots the last living specimen or a bulldozer scrapes away its last remaining habitat. Extinction is the result of the unraveling of the web of relationships and the survival pathways that sustain the species. To put it in anthropomorphic terms, the salmon are on the edge of extinction because of a loss of "ecological companionship."[13] An important insight to the cause of the unraveling of their ecological relationships can be summed up in the three words, "sense of place."

⌒⌒

Sense of place has roots in the Latin *genius loci*, which originally referred to a spirit or a divine protector of a place or community. Over time, the belief in protective spirits waned and *genius loci* became a special feeling imparted by a place. Urban planners and interior decorators use that meaning to describe a special ambiance in a city park or the interior of a house.[14] Sense of place also describes the special relationship between humans and their habitat. It is a "deepening into one's place of dwelling" that "establishes a connectedness with place based on care and respect, which is then naturally extended to all places the deeper one goes."[15] A deepening into place means going beyond the mere naming of parts of the landscape or the artifacts of human cultures. It is the striving to understand relationships among the constituents of place.[16] It is this latter meaning that is directly associated with the salmon's problem. It is a part of the salmon's problem buried deep in our culture's stories, myths, and assumptions about nature. Sense of place is submerged below the normal line of sight in the hidden part of the environmental iceberg.

Barry Lopez says that humans are confronted with two landscapes, one external and the other internal.[17] Sense of place is the interaction between them. The external landscape includes physical and biological attributes such as mountains, valleys, rivers, climate, animals, and plants. To make sense of the external landscape, we give these attributes names and then we measure or count them. But real understanding requires careful observation through which relationships among and between the living and non-living parts of the landscape are revealed. The internal landscape, the landscape of the mind, tries to understand the external landscape through knowledge of those relationships.

Our understanding of salmon for most of the twentieth century was based on numbers obtained at various times in their life history. Management success or failure was defined as either an increase or a decrease in the number of fish. But salmon are at the center of a dense web of relationships. They are consumed by at least 137 vertebrate species.[18] The ocean-derived nutrients they carry to their home rivers create a host of ecological relationships as those nutrients move through aquatic and terrestrial food webs. We cannot hope to understand and manage the salmon intelligently without understanding that web of relationships and incorporating them into policies and programs. Discovering and paying attention to those relationships are part of "deepening into a place."

Lopez's two landscapes suggest a two-part approach for this examination of sense of place and its role in salmon depletion and recovery. The two-part approach is characterized by two questions: What constitutes a place (the external landscape), and how do we sense or experience it (the internal landscape)?

What makes the Pacific Northwest unique among the bioregions of North America? For me, the answer to that question begins with the region's geology. Extinct volcanoes and others not so extinct remind us that this is one of North America's most geologically active regions. The Coastal, Olympic, Blue, Klamath, and Cascade Mountain ranges, as well as the western tip of the Rocky Mountains, are prominent geologic features. Between and among the mountains are valleys, plateaus, and plains. Water moving from mountains through the valleys and into the ocean carved—and continues to carve—out river basins. Atmospheric processes interact with physical geography, producing a mix of climates: heavy rainfall on the western side of the mountains and deserts on the eastern side. Temperature gradients from the valley floor to the mountaintops are visible in shifting snow levels. All of this has created a variety of micro climates and a patchwork of local habitats for native plants and animals, some of which— the salmon runs and the coastal rain forests—attained phenomenal levels of productivity.

Humans play an important role in shaping and defining place. They are as much a part of place as the mountains, Douglas-fir, rivers, and salmon. The interaction between humans and place is mediated by culture. I copied the following statement from a display in the Native American section of the Provincial Museum in Victoria, British Columbia, in March 1998.

In order to survive, humans must provide for their material, emotional and intellectual needs. These are satisfied by a culture, a complex system that includes tools, language and beliefs. Cultures vary because they must be compatible with their supporting environments. Thus different climate, terrain and sources of food evolve different cultural responses.

The statement sums up the relationship between place and Native American culture.

Native American culture coevolved with the fauna and flora and the physical landscape of the Pacific Northwest as it emerged from under the Wisconsin Ice ten thousand years ago. The myths, tools, and rituals of their culture were adapted to the opportunities and constraints in the local habitat. Strong feedback loops guided the evolving culture into a sustainable relationship with the salmon and other resources. If a tribe overharvested the local run of salmon or the local clam beds, they quickly paid for the transgression with a reduced supply of food. Native Americans had strong ties to their habitat. They, like the salmon, had a strong relationship to place.

Euro-Americans exploited—and continue to exploit—natural resources without the constraints of strong feedback loops. Distant markets determined the rate of exploitation without regard to the status of the resource. Many newcomers believed that the fish they were catching and the trees they were cutting were inexhaustible. The drive to satisfy markets (by converting resources to cash) and the seemingly endless supply of those resources dissolved the feedback loops that might have restrained exploitation. Markets were such a dominant part of the Euro-American culture that feeding the market's insatiable appetite for natural resources overrode stewardship and conservation. Native Americans were tied to place; Euro-Americans were tied to markets.

Place is more than a collection of things: mountains, rivers, valleys, native plants and animals, and the artifacts of human cultures. Events, especially cyclic or recurring events, are also an important constituent of place; in other words, places don't just exist, "they happen."[19] A recurring event can be so unique, so connected to the landscape, the flora and fauna and human cultures, that place and event merge, becoming inseparable. In the Pacific Northwest, the annual return of the salmon was that kind of place-defining event. Timothy Egan stated it nicely when he said, "The Pacific Northwest is simply this: wherever the salmon can get to."[20]

Prior to the arrival of Euro-Americans, the return of the salmon was an ecological, spiritual, and economic event of monumental importance. It generated a sea of relationships that tied together the region's geology, climate, flora, fauna, and the aquatic and terrestrial ecosystems. For thousands of years, the salmon's return was an important spiritual and economic event for Native Americans. The resulting relationships between the salmon and their ecosystems were so extensive and important that ecologists gave salmon the status of keystone species.[21] Once again, Timothy Egan captured the importance of salmon when he said, "In the Northwest, a river without salmon is a body without a soul."[22]

Today, the Pacific salmon are extinct in 40 percent of their historical range and, where they still exist, the runs are a small fraction of their historical size.[23] Many rivers are so degraded that the salmon cannot complete their freshwater life histories. To compensate for lack of habitat protection, salmon are taken from rivers and placed in fish factories where their offspring survive in the artificial environment of concrete ponds. Domesticated salmon—those that adapt to the artificial habitat of the fish factory—are afflicted with a condition called "ecological placelessness."[24] The domesticated factory salmon are no longer adapted to their natural habitat. They are aliens in their home rivers.

The return of factory-produced salmon to rivers where they can no longer complete their life cycle or do so only with difficulty raises the questions: Is the return of the salmon still the place-defining event for the Northwest? Have the dammed, dredged, polluted, dewatered, silted, heated rivers become stage props where fishermen and the engineered[25] facsimiles of wild salmon continue to play out their roles reenacting an event that has become a hollow shell of its historical predecessor? The engineered facsimiles still return to fish factories and, in some rivers, it's possible to find a few wild salmon. But, because of their diminished size, the runs of "ecologically placeless" salmon no longer carry the ecological and economic importance that they did for several thousand years. Even the historical flow of nutrients from ocean to freshwater has, in some years, been reversed by the massive use of fish factories. When money became the overarching metaphor for natural resources; when stewardship took a back seat to the drive to satisfy markets; and when feedback loops were stretched to the breaking point, the region started down the path of the salmon's impoverishment. All of this was aided in no small measure by a parallel erosion in our sense of place.

In my opinion, a consequence of our descent into placelessness has been the gradual shift in the place-defining event away from the return of the salmon and toward the functioning of the market. Market activities reach into every corner of the bioregion and they interact with all the other elements of place: geology, climate, fauna and flora, and culture. Travel across the region and you will see evidence of the market's invisible hand and its penetration into every facet of the landscape. Nearly all the infrastructure—roads, bridges, dams, shopping malls, airports, and cities—is directly or indirectly tied to the market. Even schools have taken on the role of providing graduates for the "job market." Although there are some biologists who will claim otherwise, the real purpose of hatcheries is not the conservation of fish. A fundamental purpose of modern hatcheries is to serve the markets dependent on salmon: the commercial fisheries that feed the markets for fresh and processed salmon and the recreational fisheries that support a large market for motels, boat manufacturers, guides, marinas, and those who manufacture and sell fishing equipment. Hatcheries are the servants of the markets; the same markets that played an instrumental role in the fall of our regional icon.

Markets seek economic efficiency by homogenizing everything, including cultures and ecosystems. Visit a Walmart or a McDonald's in Portland, Maine, or Portland, Oregon, and there is little to suggest they are in different places. Visit a salmon factory in Indiana, Michigan, Oregon, or Washington and the only way to tell what state you are in is by the state patch on the uniforms of the hatchery workers.

From the salmon's perspective, the Pacific Northwest is a diverse patchwork of micro habitats. Each variation in habitat presents the salmon with a different survival challenge. During their long history, wild salmon evolved unique life histories to deal with those challenges. Life history diversity is an important part of the salmon's evolutionary legacy. Consistent with the markets they serve, fish factories strip this away to achieve an efficient factory-like operation. I'm not building an argument to get rid of markets; they are here to stay, at least for the foreseeable future. I do believe that markets should not be allowed to seek their homogenized economic efficiency at the cost of place and the things, like salmon, that constitute place. The market's invisible hand, especially in its global reach, should not be allowed to dissolve the feedback loops that connect people to place.

A healthy sense of place would embrace ecological and cultural diversity and would, as Wes Jackson observed, cause us to approach problems through "elegant solutions predicated on the uniqueness of each place."[26] As the market assumes more and more of the role as the place-defining event, problems will be approached through simple but profitable solutions predicated on the homogenization of place. Freewheeling market ideology is antithetical to a healthy sense of place and the diversity of attributes that constitute place.[27] From time to time I hear well-meaning environmentalists declare that we need market solutions to environmental problems. Markets may have a role, but those who propose market solutions have an obligation to show both that they understand that markets are part of the problem and that they can control the market's negative effects while using them for useful purposes.

The dominance of the market is not just an environmental problem. On September 11, 2001, a group of terrorists launched a major attack on America, killing more than three thousand of our fellow citizens. The attack united all Americans, gave them a sense of duty and obligation to their country and a renewed patriotism. A united nation stood behind the president ready to act and to follow his direction. What direction did he give? He told us to go shopping. Have we really gone that far? Has our patriotic duty as citizens during a time of crisis been reduced to a market transaction?

The constituents of place tell us little about how we actually experience and think about it—the sense in "sense of place." How we perceive place involves what Barry Lopez calls the internal landscape, the landscape of the mind. But the perception of place also involves our basic sensory attributes—the ability to see, feel, taste, and hear a place. Humans tend to transfer their perception of place to other species and believe there is one natural world and it is the one that we experience. However, species vary in their sensory attributes, which means there are many natural worlds, each defined by the sensory attributes of the species experiencing it.

Jakob von Uexkull wrote a fascinating paper, a "ramble," as he described it, through the diverse worlds experienced by different species including an interesting account of the tick's sense of place. Ticks are blind and live in a world defined by three sensory attributes. They can detect light, smell butyric acid, and sense warmth. The tick's photosensitive skin stimulates it to climb upward on shrubs or small trees. It rests there until it detects the odor of butyric acid, which is common on the skin of mammals. Having detected the presence of its prey, it falls from the shrub, hopefully landing on the source of the odor. The tick knows it has found food if it senses the mammal's warmth, and this stimulates it to dig its head in and gorge itself on blood.[28] Although we can try to describe the worlds of other species based on our knowledge of their sensory attributes, we can only experience the world through our own ability to see, feel, taste, hear, and smell. Compared to the tick, humans have the potential to perceive and experience a world that abounds in rich detail. We have that potential, but human cultures can be powerful filters that enhance or diminish our connection to place.

Imagine you are on a large Polynesian canoe in the Pacific Ocean with no land in sight. The blue waters are in constant motion and, as far as you can see in every direction, the sea is a vast and empty sameness. There are no landmarks and no identifiable features in the seascape to help you determine where you are. The abstract lines of longitude and latitude that your culture has drawn across the earth are invisible and useless without the aid of a global positioning system (GPS), maps, or other navigational aids. But a Polynesian navigator looking at the same seascape does not see a blank sameness, but several different kinds of waves, each telling him something about his location. The location of evening stars, the presence of specific kinds of clouds, the movement of birds, and the kinds of waves, their speed and direction, are useful clues used by Polynesian navigators to determine their location. Their perception of place is so deep that in one case, after being blown off course by a storm that lasted for three days, a Polynesian navigator was able to determine where he was and continue his journey.[29] Unencumbered by our cultural filters, the Polynesian navigator sees a seascape full of information, full of things that are invisible to folks from other cultures.

Cultural filters are rooted in the beliefs, myths, and stories about how nature works—the hidden part of the environmental iceberg. Those beliefs and myths are buried so deep in our culture that they are rarely called into question, yet they exert a strong influence on our sense of place. For Euro-Americans, our current set of cultural filters emerged in the seventeenth century as an offshoot of Newtonian Mechanics. Newton described a natural world that functioned with the regularity and precision of a machine. That vision gave rise to the mechanical worldview, which has dominated European and Euro-American cultures ever since. The root metaphor of this worldview is the machine.[30] When ecosystems are viewed through the mechanistic filter, they become organic machines that function according to a set of fixed laws and simple causal relationships among the parts. This renders history and especially evolutionary history irrelevant, because the operation of a machine is the same day after day. The complex relationships among the inhabitants of ecosystems are replaced by the simple and predictable relationships of the machine's parts. Parts in the machine do not interact in ways that cause them to change, grow, or coevolve in response to each other.[31]

The philosopher and mathematician Alfred North Whitehead complained that the mechanistic worldview filters and deems irrelevant most everything except those things that can be measured and quantified. It reduces nature and human experience of it "to clear-cut definite things with clear-cut definite relations." Nature viewed through the mechanistic filter consists of little more than the things that can be quantified—the number of trees and salmon, gallons of water, acres of grass, or miles of stream. While Whitehead acknowledged the technological benefits that the mechanistic worldview has produced, he complained that it created a willful blindness to the many important attributes of nature.[32]

The machine metaphor for nature and the money metaphor for natural resources are complementary. Nature is a machine producing commodities, which are then converted to dollars through markets. Whitehead's use of the term "clear-cut" was appropriate. The mechanistic worldview clear-cuts nature of all its internal relationships, its coevolving and emergent qualities, and the enchantment that should come from a personal experience with the richly diverse mantle of life that covers the earth. It clear-cuts anything that can't be defined by a number, or anything that can't be converted to cash. There is no room for Gary Nabhan's "ecological companionship" in the mechanistic worldview.

There is no doubt that in medicine and technology the mechanistic worldview has produced tremendous benefits for humans.[33] Advances in medical technology, communication systems, computer technology, water and sewage systems, and intensive agriculture were facilitated by the machine metaphor, but it also converted rivers to hydroelectric machines, transportation systems, and sewers. It diverted water from rivers to irrigate farms and it created the belief that salmon production could be industrialized in fish factories. The fish factory and the machine metaphor are a perfect match. The mechanistic worldview reduced salmon-sustaining ecosystems to an industrial process and rivers to simple conduits whose only function was to carry artificially propagated salmon to the sea. The mechanistic worldview still has a powerful grip on salmon management and restoration programs in spite of a growing scientific understanding that the picture of ecosystems created by the machine metaphor is seriously flawed.

I have walked through mature forests with industrial foresters who could only see logs, board feet, and cash. I spent one afternoon with a forester who found it incomprehensible that the federal government would not permit logging in the Olympic National Park. According to his myopic view, the old-growth firs and cedars in the park were simply being wasted. I know salmon managers whose vision of Oregon and Washington rivers and their salmon runs is limited to "fish in the boat." They care about the forests and salmon, but they also believe that forests and rivers are warehouses full of commodities that they have an obligation to harvest in a way that maximizes their conversion to cash. Cultural filters cause them to see waste in a tree that dies and decomposes on the forest floor or in salmon that escape the fishery in excess of the minimum number needed to seed the habitat. Decomposing trees and salmon contribute to the health of the ecosystem by releasing nutrients to food webs, but those benefits are rendered irrelevant and invisible when ecosystems are thought of as machines. In spite of our sensory richness there are humans whose cultural filters are so myopic that they walk the landscape like large ticks with a greatly impoverished sense of place. The cultural filters that narrow our perception of the landscape, that distort our sense of place by clear-cutting out of our vision everything except the number of commodities or that limit all value to cash, cause me to worry about the fate of the salmon and the fate of this place.

So far this discussion of place has focused on the human sense of place. It's time to consider place from the salmon's perspective. After migrating hundreds, if not thousands, of miles in the ocean, wild salmon return to their home stream. They make their often heroic migration upstream, and spawn in the same place where they were born. This attachment to place is one of the salmon's important biological attributes because it is the source of the species' biological diversity. The drive to return to the place of their birth isolated the individual breeding populations and, over thousands of years, those populations adapted to the environmental conditions in their home stream. This created a diversity of genetically distinct populations. Their genetic diversity combined with the complex mosaic of habitats encountered during their long migrations resulted in a rich diversity of life histories, which enhanced the salmon's survival, giving them the resilience needed to cope with environmental fluctuations.

This chain of connected attributes—attachment to place, genetic and life history diversity, and resilience—was largely responsible for the highly productive wild salmon populations present when Euro-Americans arrived in the Pacific Northwest. The wild salmon's attachment to place is the wellspring of much that makes them a treasure deserving of real stewardship.

How have the salmon management institutions treated the salmon's attachment to place? Have they developed place-based management policies and programs? Let's look at those questions by examining the salmon manager's two principal tools: fish factories and harvest regulation.

Fish factories have always been used as a substitute for place. Salmon managers traded habitat for hatcheries. The availability of factory-produced salmon that could be planted in degraded rivers weakened the incentive to vigorously protect the salmon's habitat. In their operation, fish factories ignored the salmon's attachment to place. Salmon heading for other parts of a watershed or for entirely different rivers are trapped and subjected to the environmental conditions in the hatchery. The environmental conditions in the salmon's natal stream, the conditions they are adapted to, do not enter into the factory-like operation of the hatchery. In this regard, hatcheries are more closely related to cattle feedlots than to the river where the factory-produced fish are dumped. Salmon brought into the hatchery become domesticated; they adapt to the artificial habitat of the fish factory as they once adapted to the environmental conditions of their home stream. When the domesticated salmon stray into the natural spawning grounds and spawn, their offspring find it difficult to survive. Domesticated salmon are ecologically placeless and science tells us that domestication, which is demonstrated by lower survival after release from the factory, begins immediately after wild salmon are taken into a hatchery.[34]

Two events in the closing decades of the nineteenth century and the opening decades of the twentieth century contributed to a placeless approach to harvest management. First, switching from sail to gasoline engines to power their vessels gave fishermen the ability to move from the river to the ocean, where they could intercept salmon before they entered their home river. Harvest was no longer on local stocks after they entered in their home streams, but on mixed aggregates of stocks from several rivers. Salmon targeted in the ocean fisheries might be caught hundreds of miles from their home river.

The rapid growth in ocean salmon fisheries coincided with the creation of resource management agencies staffed with technical experts.[35] Centralized decision making in these bureaucracies led to uniform harvest regulations over large oceanic areas. Uniform regulations were applied to mixed aggregates of stocks regardless of the productivity and status of the individual populations.[36] The individual salmon populations in those stock aggregates received little attention because they were rendered invisible by Euro-American cultural filters. W. F. Thompson summed up the consequences of this shift in management perspective in a paper published in 1965, just before he died. After describing the importance of understanding how salmon populations are adapted to their local habitats, Thompson described the consequences of a placeless harvest management that ignored the importance of the individual populations:

> But we do not know much about these independent, subspecific groups
> of salmon segregated during spawning, and so we do not know just how
> to conserve the numerous kinds that exist. In our fisheries, we have been
> accustomed to dealing with mixtures of many of these units, although each
> has its own particular requirements. ...We can only moderate our ruthless
> fishery, blindly and in partial fashion; we cannot avoid its effects completely.
> ... knowing only that our total catches diminish, as one by one small
> populations disappear unnoticed from the greater mixtures which we fish.[37]

Harvesting mixed stocks of salmon in the ocean and trying to compensate for the resulting diminished supply of fish by making ecologically placeless animals in fish factories is placeless management. What did placeless management accomplish? Pacific salmon are now extinct in at least 40 percent of their historic range and the salmon in most of the remaining range are under the protection of the federal ESA. Management that ignores one of the salmon's important biological attributes is bound to fail—and it has failed.

It is nearly noon when I finish inspecting the stream. The loggers have complied with the rules and left a thin band of skinny alders along its bank. Surrounded by the treeless clear-cut, the alders give, at best, flimsy protection to the stream. Without the surrounding forest to buffer the alders from strong winds, they will most likely be blown down this winter. Similar to the abstract lines that guide the jumbo jet, the lines that enclose the protected riparian areas are artificial human constructs. The rules regarding stream protection are the result of bargaining, hard-ball politics, and compromise covered with a thin veneer of science. Like the false fronts of western towns used in movie sets, these thin charades of riparian forests only fool those who find it convenient to be fooled.

The boundaries are inadequate and bear little relation to the role of the riparian trees in the formation and maintenance of stream habitat.

I hike back to the truck and look for the winter wren, but it is gone, so I drive off to look at another clear-cut. There is always another clear-cut.

Side Channel 1
Finding an Old Friend
a Long Way from Home

It is the contention of these essays that the organism is most interesting and healthiest when it is exposed to an environment fundamentally in harmony with the environments that, in the evolutionary process, were responsible for producing the bundle of adaptations that the individual represents.

—Daniel Kozlovsky[1]

The late morning sun was gradually warming the frosty air, while the river, smooth as glass, slipped silently through a landscape dressed in fall colors. When I stopped to pull off my jacket, I saw her. She was lying under a hickory tree on the bank of a small stream a few feet from where it joined the river. Her eyes, which once blazed with a fierce determination, were dull and lifeless. Her once bright silver sides were tarnished to a dull gray.

In the states of Oregon or Washington, where I have lived for the past forty-five years, it's not unusual to see dead salmon in and around the rivers in the fall. But this was the St. Joseph River near South Bend, Indiana. This salmon was twenty-three hundred miles from home. Picture an elephant walking through an Iowa cornfield or a clown fish from Australia's Great Barrier Reef in a Montana trout stream and you get an idea how strange this salmon looked on the banks of a stream in an oak-hickory forest. I bent over the dead salmon and said: "You're a long way from home."

Over the next several days I saw more salmon. They were in a strange place, in a strange river; nevertheless they faithfully followed the rules of behavior recorded in their genes. Several of the salmon were working hard against the river bottom trying to build redds. I knew it was a futile effort. Salmon eggs need to incubate in redds where the gravel is clean and porous, so oxygen-rich water can percolate through to the embryos. I could see that the gravel was plugged with suffocating silt and sand. Few, if any, of the eggs would survive. A couple days later, I was back watching the salmon trying to spawn. A biologist from the Indiana Department of Natural Resources came by and confirmed that there was little natural reproduction of salmon in the river. In other words, the St. Joe was not really a new home for the salmon. They were here only because humans take them from the river, where they have little or no chance of completing their life cycle, and artificially propagate them in fish factories. The salmon I saw in the St. Joe River were the ultimate in ecological placelessness.

I applaud the states of Indiana and Michigan, the Friends of the St. Joseph River, and others that have helped clean up the river. When I was growing up here several decades ago, the river was largely used as a sewer. As I watched fishermen enjoying the opportunity to catch a salmon, I found it hard to believe this was the same river I knew as a boy. It appeared I was witnessing an environmental success story, but as I sat on the riverbank watching the salmon going through their futile spawning ritual, a troubling question nagged at me. Should we take these magnificent animals from their home streams in the Pacific Northwest and put them in a river where they have little hope of completing their life cycle? Do we have to plant salmon in every river the invisible hand of the market points to?

A few days later that question was still bothering me when I visited the local salmon hatchery. It was exactly like so many others I have seen, right down to the displays and brochures that praise the virtues of fish factories. As I walked around the concrete raceways, I was thinking that hatcheries have a lot in common with big box stores like Walmart or Costco and restaurants like McDonalds or Applebee's. They all show the effects of homogenization—an affliction common to those things that primarily serve markets. Not far from the hatchery I watched three men fishing for salmon. Their gear—waders, rods, reels, lures, and clothes—matched what I would see fishermen using in any stream in the Pacific Northwest. It was all the same except for the natural setting. I could see nothing in the natural setting that said, "this is a salmon place."

Pacific salmon have a heroic life story. The word salmon invokes the image of large silver fish struggling to ascend wild rivers where they spawn and die. Their bodies are a feast that the whole ecosystem partakes in, including the next generation of salmon. Nutrients carried from the ocean flow into aquatic and terrestrial food webs where they radiate through the ecosystems. The salmon give strength to Douglas-fir, cedar, and other riparian trees, and they soar in the bodies of eagles.

Placing salmon in ecosystems where their survival depends on a fish factory robs them of their story. The river becomes a stage prop where fishermen and fish act out their parts. The salmon become widgets produced in a factory for a market selling virtual experiences. This also occurs in the salmon's real home. In the Pacific Northwest, rivers are blocked by dams, pumped dry by irrigation, subjected to pollution and massive habitat degradation. In those rivers salmon also depend on fish factories. What I was witnessing here in Indiana and in Oregon was resource management that ignores the importance of place. It is management without stewardship.

Placeless management is essentially an attempt to rewrite the salmon's story, which was so painstaking crafted over millions of years of evolutionary history, before we have assembled, read, or understood a complete version of the original. The story is being revised to fit the universal, homogenized plot of the market economy. I came away from the experience in Indiana with a heightened sense that the current approach to salmon management is on the wrong path. Is Indiana the same as Oregon? Is Oregon the same as Indiana? Placeless, market-driven management is saying yes to those questions.

I went back to see the Chinook salmon before returning to Oregon. She was still there under the hickory tree. I bent over and said, with a deep sense of guilt for what my species has done to hers, "Goodbye old friend."

Chapter 2—Salmon Stories

The irony here is that we profess to be a knowledge-based culture—modern society claims to have abandoned myth for the safety of solid science. This itself may be our greatest cultural myth.

—William Rees[1]

Each hypothesis is the nucleus of a story, and gradually the stories accumulate and link to each other, merging into a master narrative that explains what happened.

Smart people in every culture proceed in this manner to reveal what they can of the mysteries of their day, or at least the mysteries they feel motivated and permitted to pursue.

—William deBuys[2]

Humans have been telling salmon stories for thousands of years. Those stories explain the source of the salmon's abundance, define the terms of the human-salmon relationship, and set the rules that govern our behavior toward these magnificent animals. A culture's stories can lead to long-term survival and prosperity or to crisis.[3] Thomas Berry tells us that our biggest crises occur when our stories no longer help preserve the things we value.[4] Our salmon story has failed to preserve what we value, our regional icon, and given rise to a crisis. In an article by Scott Learn, the Portland *Oregonian* (May 12, 2008) once again reminded its readers that the crisis has real consequences. The article, "Salmon Closure Hits Winchester Bay Hard," described how the continuing depressed status of salmon runs is a source of hardship for coastal communities—a crisis for both salmon and salmon people.

When I think about the differences in the stories that guided the behavior of Native Americans and Euro-Americans toward the salmon, two very different images come to mind. One is of a whole village turning out to perform an elaborate ceremony in honor of the year's first salmon. They treat this salmon with great deference. After the salmon is eaten, its bones are carefully returned to the river.[5]

The other image is of a cannery worker shoveling rotting salmon carcasses off the dock and into the river. Before the collapse of the large salmon runs, fishermen often caught more salmon than a cannery could process, leaving many on the dock to spoil. To achieve maximum economic efficiency, canneries often bought more salmon than they could process to ensure that there were no lapses in the supply of fish to be canned. The salmon story that led to the latter image is the subject of this chapter and, to begin, we need to go back to the dawn of the industrial salmon fishery in the closing decades of the nineteenth century.

When Euro-Americans arrived in the Pacific Northwest, they found an entrepreneur's dream. The rivers were full of large silver fish with high-quality flesh that retained its quality for years in the can. The salmon literally swam into the fishermen's nets. The timing of the runs varied little from year to year, so fishermen and cannery operators knew when to have their machines serviced, their boats and nets repaired, and the supply of cans on hand. The uniform size of the salmon quickly led to mechanical butchering machines, making it possible to convert the salmon to cash at a faster and faster rate. Salmon easily made the transition from the subsistence economy of the Native Americans to the industrial economy of the newcomers.

But the cannery owners and managers were not satisfied with the wild salmon's natural productivity. Even though they had no real understanding of the ecological processes that sustained the salmon's massive abundance, they believed that they could improve on nature if they manipulated and controlled salmon production. In the late nineteenth century, the nation was rapidly converting to an industrial economy, so the salmon managers and cannery operators naturally looked in that direction for a way to enhance the supply of fish. Since the canneries had already industrialized the utilization of adult salmon, the managers and entrepreneurs turned their attention to the other end of the salmon's life cycle. They industrialized the salmon's reproduction through the use of hatcheries. To underscore their industrial genesis, the early hatcheries were called fish factories. Salmon factories gave humans control over production—or so they thought—and armed with that control, managers believed they could break through the natural limits on salmon abundance and increase the supply of these valuable fish the same way agriculture increased the production of food crops and animals.[6]

In the 1870s, salmon managers, cannery operators, fishermen, and politicians enthusiastically turned to salmon factories to increase the supply of fish. The belief that fish factories would produce salmon in such abundance that it would exceed the natural, pristine productivity of rivers had no factual or scientific support. However, that belief was consistent with the prevailing ideology, which had deep theological, economic, and political roots. An ideology with theological, economic, and political roots didn't need factual support.

In his landmark paper, "The Historical Roots of our Ecological Crisis," historian Lynn White discussed the theological roots.[7] White said that Christianity created a major shift in attitudes towards nature when it separated humans from the rest of the biotic community. One consequence of this separation was the belief that humans must dominate nature and exploit it for their benefit. Nature was no longer a community of beings exchanging gifts, but a warehouse of commodities, and only humans held the key to the door. Nonhumans that trespassed into the warehouse were shot. Eagles, seals, and sea lions that dared steal salmon from the warehouse were killed in the tens of thousands. Taboos against human overexploitation and waste of salmon were considered quaint reminders of an older culture. Twelve years after the first salmon factory began operation, George Goode of the U. S. Fish Commission mirrored the belief in human dominance over nature when he relished the idea that hatcheries would bring salmon production under the complete control of humans.[8]

White's description of Christian theology and its effect on the environment is not universal among churches in the Pacific Northwest, especially in recent years. In 1990, Pope John Paul II released his environmental message with the title "The Ecological Crisis: A Common Responsibility." The Catholic bishops of the Pacific Northwest followed up on the Pope's message with "The Columbia River Watershed: Caring for Creation and the Common Good." The Oregon Ecumenical Ministries has an environmental ministries program.[9] The ecumenical recognition of our obligation to care for creation reflects the public's growing concern for the environment. So far these changes have not reached into the government agencies and changed their approach to salmon management.

Fish factories were consistent with prevailing economic beliefs in two ways. First, hatcheries underwrote the belief that entrepreneurs should have laissez-faire access to natural resources. If hatcheries could maintain the supply of fish, if concrete ponds could replace rivers, then the watershed's other resources could be developed and exploited with minimal regulation while the supply of salmon was maintained. Fish factories became the silent partner that facilitated the economic development and degradation of the region's rivers. They were the key to the minimal regulation of irrigation withdrawals, grazing, logging, dams, hydroelectric production, industrial and urban development, and pollution. Second, hatcheries were an example of the theoretical concept of substitutability, which underlays the belief that perpetual economic growth was desirable and possible.[10] The fish factory substituted an industrial process for the ecological processes that made up natural, salmon-sustaining ecosystems.

Finally, hatcheries were political tools. They gave politicians an easy way to avoid conflicts between the salmon-canning industry and other economic activities that destroyed or degraded aquatic habitat. Everyone benefited: salmon managers were given fat budgets to run the fish factories, politicians were given credit for solving a problem, and developers were given access to rivers with little regulation. Everyone benefited except the salmon and ultimately the fishermen and their communities.

A bargain was struck. Salmon factories were traded for habitat. Over the next century the bargain became a comfortable myth and a central part of our salmon story. It was so comfortable that no one bothered to notice that hatcheries weren't living up to their end of the deal. When managers and politicians did not hold fish factories accountable, they changed them from a new management tool that needed evaluation to a comfortable myth. The lack of accountability was a key element in this transformation because myths can only survive if they remain unexamined.[11]

Once they escaped the penetrating light of accountability and slipped into the darkness of myth, hatchery programs multiplied and became the basis for a simple, industrial model of the salmon's production system. In this model, hatcheries sent juvenile salmon to the ocean where they grew and were caught by sport and commercial fishermen. Managers regulated harvest so enough salmon returned to the hatchery to start the cycle over. This simple model is the core of our salmon story and it has been the underlying foundation of management for over a century. It is composed of two primary activities: the operation of fish factories and the regulation of harvest. Since those two activities are also the main conduits channeling funds into management agencies (licenses and taxes related to harvest and budgets to operate hatcheries), it is not surprising that they are protected from close scrutiny.

Consistent with the belief that habitat could be traded for hatcheries, the river was not treated as an ecosystem, but as a simple conduit to carry factory-produced fish to the sea. Rivers were reduced to salmon "highways" and, like real highways, they needed to be clean and unobstructed.[12] This view of rivers led to their wholesale "cleaning" in the 1950s, '60s, and '70s, which simplified and degraded natural salmon habitat. Seeing rivers as highways and salmon as traffic shows how the industrialization of salmon production and the salmon story influenced the way managers visualized the salmon's ecosystems. It is an example of one of those cultural filters discussed in the previous chapter.

This salmon story is heading for a sad ending because it has a fatal flaw. It assumes hatchery fish do not require healthy river habitat and that incubation trays and concrete raceways can replace the salmon's natural spawning and rearing habitat. The new salmon story rendered the ecology of the river and salmon habitat irrelevant. Accordingly, the first surveys of salmon habitat in the Columbia Basin did not take place until the 1930s, nearly seventy years after fish factories began operating. Then it was another fifty years before that information was actually used.[13] In recent decades there has been a lot of talk about the importance of habitat and it seems that's all it is—a lot of talk. All we need do is "follow the money" to see salmon managers' priorities. The federal Government Accounting Office (GAO) did just that for the Columbia Basin. In its review of the salmon recovery programs in the Columbia Basin the GAO estimated that up to 1980 only 1 percent of the expenditures on salmon recovery went to habitat restoration. Fish factories, on the other hand, consumed about 45 percent of the budget. Between 1980 and 1990, with the threat of listings under the federal ESA, habitat restoration and protection received more attention. Habitat's share of the budget was boosted to about 6 percent. Despite nice-sounding rhetoric to the contrary, our salmon story has not given habitat a high priority.[14]

To achieve their political purpose and reduce or prevent conflict between salmon and the economic development of rivers, hatcheries had to render natural habitat interchangeable with concrete ponds. Proponents of economic development that degraded salmon habitat shifted the public's attention away from dewatered, heated, silted, poisoned, and dammed rivers, by pointing to fish factories and saying, "don't worry, hatcheries will replace habitat and maintain the supply of fish."

But it wasn't just physical habitat that fell victim to the enthusiastic endorsement of hatcheries. Fish factories also eliminated the need to worry about all the messy ecological relationships that linked salmon to their ecosystems. For example, salmon eggs and flesh are food for at least 137 species.[15] When salmon are taken from the stream and placed in the fish factory they are removed from the food webs of those 137 species, and the relationship between salmon and those species is severed.[16]

The complex and seemingly chaotic ecological relationships that make up the salmon-sustaining ecosystems were replaced by the hatchery with its neatly manicured lawns and the factory-like functioning of its ponds and other infrastructure. Fish factories reek of order, efficiency, and control. The appearance of order and efficiency hid the ecological disaster that fish factories were facilitating and reassured the general public that salmon managers had it all under control and there was no need to worry. The posters and brochures that described the salmon factory in glowing terms conveniently left out any discussion of the bargain and the fact that hatcheries were failing to hold up their end of the deal.

Another consequence of the industrial substitute for salmon-sustaining ecosystems is the failure to appreciate the importance of salmon life histories— the sequence of events in a salmon's life from birth to death. W. F. Thompson used a chain as a metaphor for life history. He said the salmon's life history is a chain of habitats where the fish complete the sequence of events or activities (spawning, rearing, and hiding from predators) that compose its life history.[17] This definition inextricably ties life histories to habitat. Each link in the chain binds an event in the salmon's life history to a physical attribute of the stream, river, estuary, or ocean. It's impossible to understand the importance of habitat attributes without considering the way they fit into the salmon's life history. On the other hand, understanding life history requires knowledge of the chain of habitats where the events of that life history are played out.

Thompson went on to say that a salmon population may be composed of several life history-habitat chains. These alternative life histories are an important part of the salmon's overall biodiversity. Life history diversity is the salmon's way of solving the problem of survival in the mosaic of variable habitats and fluctuating environmental conditions that it encounters from the headwater spawning areas to the ocean.[18] Life history diversity can be visualized as a network of spatial-temporal pathways through the river, estuary, and ocean. As habitats change or environmental conditions fluctuate, the survival value of specific pathways will also change.[19] This web of spatial-temporal pathways is how salmon avoid placing all their eggs in one survival basket. However, in the industrial model of the salmon's production system, life history diversity is no longer an important evolutionary legacy; it is an impediment to the economic efficiency of the fish factory. Fish factories are economically efficient when all the salmon do the same thing at the same time, when they all follow a single life history. So along with habitat and ecological relationships, fish factories rendered the salmon's biodiversity irrelevant.

⌇⌇

The ways in which salmon life histories and habitat are bound together reminds me of Bud Schmitt, a very remarkable man. Mike McHenry, a fishery biologist at the Lower Elwha Klallam Tribe, introduced me to Bud. Afterward, I stopped by Bud's place three or four times. Bud lived at the mouth of Whiskey Creek, a small stream flowing north into the Strait of Juan de Fuca west of Port Angeles. He told me he had lived at the mouth of Whiskey Creek nearly all his life and had grown up there during the Great Depression. One day as we talked, three or four seals were swimming near the mouth of the creek. I asked him if he had ever killed seals for a bounty when he was young. He said he had, but that he would kill a seal only when he needed money to buy bullets. During the Depression, his family largely lived off the land. The bullets he bought with the bounty money kept them supplied with wild meat. As long as he left some seals unharmed, he was assured of future bounty money and meat on the dinner table. When he told me this, I realized that he had a strong connection to the place where he lived. On one of my visits, he told me a story about the coho salmon in Whiskey Creek, and it was that story that made Bud so remarkable.

"Whiskey Creek is unique," he said. "From late spring through summer, a gravel bar forms at the mouth of the stream." This is a natural phenomenon in some streams. Sediment accumulates at the mouth of a stream due to wave action and forms a temporary and leaky sediment dam. I had seen bar-bound streams on Oregon's south coast, but as far as I know, it is not common in streams discharging into the Strait of Juan de Fuca. The bar at the mouth of Whiskey Creek temporarily blocked juvenile coho salmon from entering salt water. They reared through the summer in the pool of impounded fresh water. Fall storms breached the bar allowing the coho to complete their migration to sea. Then he told me Whiskey Creek's unique physical habitat had created a unique salmon. Without using the exact words, he was telling me that the coho in Whiskey Creek had a unique life history, adapted to the stream's unique habitat.

"Things started going downhill after they logged the high bench between the Olympics and the lower reaches of the creek," he said. The bench was a wet area that supplied enough water to the stream to support the year-long residence of the juvenile coho. When the bench was logged, summer flows declined and so did the coho run. Then a young biologist from Washington State visited Bud and said that he was going to enhance Whiskey Creek's coho salmon by planting hatchery fish. The fish would come from a distant fish factory. Now this is the really remarkable part of the Bud's story. Bud said he told the biologist that his plan wouldn't work unless he used coho salmon from Whiskey Creek, because they were used to the way it became bar bound each year. He told the biologist the same story he told me, but fish from another river were planted anyway. They were planted for several years. The run didn't improve, the planting stopped, and the young biologist didn't return.

Over several years of close observation and with a deep sense of place, Bud had acquired more practical knowledge about the importance of the salmon's life history-habitat relationship than the biologist had learned with his college education. No, that's not entirely correct. The biologist's education gave him the knowledge, but the salmon story and its industrial underpinnings rendered such considerations irrelevant. Bud's local knowledge of Whiskey Creek gave him a detailed conceptual map of his home place. His map was unique to Whiskey Creek, and it contained the information needed to recover the stream's unique coho salmon population. The biologist also had a conceptual map. It was exactly the same map for every watershed he was assigned to "enhance." His map was not derived from local knowledge. It was derived from the mythical salmon story and its fish factories. Bud had a sense of place and a deep connection to Whiskey Creek. He had the knowledge needed to "enhance" the coho run into Whisky Creek, but not the authority to use that knowledge. The biologist worked within the constraints of a conceptual foundation that had little in common with

Whiskey Creek, little in common with reality, but he had the authority to impose it onto streams like Whiskey Creek and their wild salmon.

Bud's story of Whiskey Creek is an example of salmon management that ignores the imperative of place; it's an example of management that ignores the unique attributes of salmon that connect them to place; and it's an example of the power of cultural filters that can severely distort what a biologist sees when looking at a stream. At first, I couldn't understand how such a wrong-headed approach to salmon management could persist, and then I read Rik Scarce's book, *Fishy Business*.

Scarce says that we haven't been managing the real salmon. The Euro-American salmon story created a mythological fish, a socially constructed animal capable of thriving in highly developed rivers. Our salmon story created an animal that does well in the concrete habitat of fish factories and an animal that behaves according to the biologists' models. The socially constructed salmon was not derived from biological or ecological imperatives. It is an engineered animal, whose attributes were fashioned by the market's invisible hand rather than a salmon-sustaining ecosystem. According to Scarce, "The result is an oddly commonplace salmon—a mechanical, schematic, engineered fish—to many of those who know the most about them."[20]

I agree with Rik's assessment, but I don't believe he went far enough. The Euro-American salmon story not only created a mythological animal; it broadened the myth to include the whole salmon production system. In this socially constructed production system, habitat was replaced by fish factories and the salmon's biodiversity became a detriment. The highly simplified production system let the managers believe they were in control and that control would allow them to mimic the success of modern agriculture and increase salmon production beyond natural levels. Advocates of this industrialized salmon production system believed so strongly in this myth that they superimposed it onto the real rivers (like Whiskey Creek) and their wild salmon populations with devastating results. Our salmon story created an industrial production system that has little in common with the ecosystems that sustain wild salmon. Using the story as the conceptual basis for management minimized the importance of wild salmon populations, undermined their biological diversity, and justified the exchange of habitat for fish factories. Belief in the story and its industrial production system made trading fish factories for habitat not just possible, but highly rational.

Once fish factories were adopted as a major management tool, the United States Fish Commission set a goal of bringing all production of Pacific salmon under human control.[21] In many rivers salmon management has largely achieved that goal. In the Columbia River, a river that drains a landscape the size of France, most of the salmon returning each year are produced in fish factories. Salmon production in the Columbia is largely under human control, but success in achieving that goal came at a high price for the salmon and the communities dependent on salmon fishing. The number of adult fish produced by factories, at a cost of several million dollars a year, is a small fraction of the historical wild salmon runs.

It's not uncommon for resource managers to create a simplified model of those parts of nature that they intend to manage for commodity production.[22] When the production of a resource is simplified, most, if not all, of the messy ecological complexities disappear. This then gives the appearance of control over the things that have economic value, such as logs, salmon, wheat, and hydropower. For salmon, simplification of the production system is clearly reflected in the measures management agencies use to show their accomplishments; these include the number of fish harvested, number/pounds of fish released from fish factories, angler days, economic value of the catch, licenses sold, and escapement. These data largely measure the performance of the industrial production system. They say little about the condition of watersheds and salmon-sustaining ecosystems, the status of salmon habitat, or the health of ecological relationships such as the relationship between the salmon's marine-derived nutrients and aquatic and terrestrial food webs and the status of genetic and life history diversity.[23]

Simplifying salmon production into the factory-like operation of hatcheries gives the illusion of efficiency, control, and success. Just look at all those thousands of little salmon in the hatchery raceways; it's a vision of success. In reality it is a well-crafted illusion enhanced by positive, if not always honest, spin. Within that illusion are the seeds of major intractable problems. A management approach that relies on fish factories ignores ecological relationships and biodiversity— but that doesn't mean they disappear. Damaged and degraded ecological processes and the loss of biodiversity will reassert themselves in the form of problems, the symptoms of which are depressed, depleted, and extinct salmon populations. Because they ignored the underlying cause of those symptoms, salmon managers are now ill prepared to deal with their source, so they focus on treating the symptoms—the declining numbers of salmon. Once the problem is defined in this way, its solution inevitably leads to increased output from fish factories. And that reinforces the original source of the problem.

So far in this chapter I could be faulted for giving the impression that the story that captured salmon management and has held it captive for the past one hundred and thirty years was somehow inevitable, that there wasn't a viable alternative, an alternative with a closer correspondence to reality. There were alternatives. The pioneering work of early fisheries biologists could have lead to a different story and a different approach to salmon management. Stephen Forbes advocated an ecological approach to the management of aquatic ecosystems in the late nineteenth and early twentieth centuries, and his work did influence fisheries management for a short time.[24] Also in the early decades of the twentieth century, in the Pacific Northwest, the work of Charles Gilbert, Willis Rich, and others stressed the importance of the salmon's life history-habitat relationships and an ecological model of the salmon production system.[25] No, it wasn't inevitable that fish factories and their simplified production systems would become the basis of salmon management. It was just easier, especially when biologists were faced with the juggernaut of dam building that began in the 1930s.

In their paper, "The Development of 'Population Thinking' in Fisheries Biology Between 1878 and 1930," Michael Sinclair and Per Solemdal discuss the importance of history in fisheries science. They refer to a comment by the French historian Fernand Braudel that the past devours the present; past beliefs, assumptions, and myths can prevent change even when the need for change is recognized. Sinclair and Solemdal believe that just the opposite occurs in fishery science. Scientists pay more attention to the current literature and are not greatly influenced by the past. So for science, Sinclair and Solemdal believe the present devours the past. Fishery scientists' lack of attention to history is an impediment to greater understanding.[26]

While fishery scientists may be influenced more by new information, I believe Braudel's statement is relevant to fishery management, and it helps explain why we have retained a failed management model for more than a century. In reaching this conclusion, I separate fishery science from fishery management. Fishery science includes the body of research conducted by academics and biologists in state, federal, and tribal agencies and others. Fishery management develops and implements programs and policies that govern the operation of fish factories and regulate harvest. They assist in the development and implementation of salmon recovery programs. Fishery management also includes a heavy dose of social and political considerations. It is more like the everyday life that Braudel was referring to, so for fishery management the past is more likely to devour the present. I would argue that in salmon management, past myths and assumptions—our salmon story—prevent or impede the use of new scientific information and they prevent change even when the need for change is obvious.[27]

Many individuals who view salmon management from outside the government agencies make the tacit assumption that new information obtained by fishery science is seamlessly incorporated into management's policies and programs. Unfortunately that is not the case.[28] New science is integrated into management programs slowly or sometimes not at all. For example, in the early decades of the twentieth century, fishery scientists recognized that the individual salmon stock in its home stream should be the basic unit of management and that practices such as the transfer of salmon from one stream to another threatened the integrity of the stock and should be stopped. Sixty-nine years later fishery scientists were still recommending that the practice be stopped.[29] The past had been devouring the present for those sixty-nine years.

Some who read this chapter will say: what he describes might be accurate history, but management has progressed beyond the old story. To a degree, they are right. Salmon management and recovery programs evolve and progress. Today, managers use sophisticated tools, such as computers, geographic information systems (GIS) mapping, genetic analysis, quantitative models, and so on. Newer state and tribal hatcheries and hatchery reform programs appear to be making use of new scientific information. But I would argue that those changes, while altering the margins of salmon management, have not changed the core story. Salmon managers now use sophisticated technology, but the application of that technology remains consistent with the assumptions, myths, and beliefs that have constrained management for over a century. The old story is buried deep in the culture of natural resource institutions and is still guiding decisions, still impeding the recovery of salmon and still devouring the present. In the next several paragraphs I'll describe some recent policies, decisions, and actions that show the continuing influence of the old story. I'll start with a personal experience.

Early in my career, I worked on the Rogue River as a research biologist for the Oregon Department of Fish and Wildlife (ODFW). The Rogue River drains much of Oregon's southwestern corner, including a small part of northern

California. It originates in the Oregon Cascades and flows 215 miles to the sea. Fishing for Rogue River salmon and steelhead is internationally renowned.[30] I went to work on the Rogue River because Oregon decided to trade salmon habitat for a dam and a fish factory. The decision had been made decades earlier when such trades still seemed rational. By the time the dam was actually built, fishery scientists were beginning to question the ability of hatcheries to replace lost habitat. I was asked to develop a research program capable of detecting any detrimental effects of Lost Creek Dam and Cole Rivers Hatchery on the wild spring Chinook population in the Rogue River. While I was preparing the study plan, I read Paul Reimer's newly published study of the life histories of fall Chinook salmon in Oregon's Sixes River. Paul's elegant research identified five juvenile life histories in the fall Chinook population (five of W. F. Thompson's life history-habitat chains) based on their length of residence in the river and estuary.[31] It was my introduction to the importance of life history and life history diversity in wild salmon populations.

Paul's study convinced me that I needed to describe the life histories of spring Chinook salmon in the Rogue River before going on to monitor their status after the dam was completed and the hatchery began operating. When I sent the draft plan to the local managers for review, one manager on the Rogue told me a life history study was unnecessary and a waste of money. According to him the life history of Rogue River spring Chinook was already known and the Rogue fish, like all spring Chinook, followed the same, single life history: The adults spawned in the fall in the headwaters, the juveniles emerged from the gravel the following spring and reared in the upper river for a year, then migrated to sea in the spring of their second year. Juvenile spring Chinook didn't rear in the middle or lower river. They passed through those areas using them as a simple conduit to the sea. I managed to keep the life history study in the project and the results were surprising. While in the Rogue River, juvenile spring Chinook followed eight life history pathways including extensive rearing in the lower river or estuary during the summer months—a life history that I was told did not exist. Juvenile spring Chinook used the entire river corridor for both rearing and migration. The single life history pathway that I was told *all* juvenile spring Chinook followed in the Rogue River was actually used by a small percentage of the population.[32]

At the time, I didn't understand the resistance to a life history study, nor did I understand the belief that juvenile spring Chinook followed a single life history. It was clear to me that Paul Reimer's Sixes River study had challenged that concept, but I didn't realize how long it took to incorporate new fishery science into management programs. It was a couple of decades later that I came to understand how much the simplifying lens of the salmon story not only distorted the manager's vision of salmon production systems, it distorted their thinking about the salmon's key attributes.

The strongly held belief that all spring Chinook followed the same life history pathway was consistent with and, I believe, was derived from the industrial model of the salmon's production system where, to achieve economic efficiency, all the salmon must follow a single life history. Managers superimposed this simplified salmon production system onto the real rivers and wild salmon runs, but before they did that, they had to superimpose those same simplifying assumptions onto their own thinking. The opposition to the life history study was, as Braudel suggested, an example of the past devouring the present. But that's not the end of this story.

Several years after I wrote the Rogue Research Plan, I was attending a workshop on the Willamette River. During a discussion of the Willamette's elevated temperature in the Portland area and the effect of that on juvenile salmon, a highly respected biologist from the ODFW reassured the audience that they needn't worry about spring Chinook. According to this biologist all juvenile spring Chinook in the Willamette River followed the same life history pathway. They reared in the upper river for a year, and then migrated out of the river in the spring of their second year. All of the juvenile spring Chinook passed through the Portland area early in the year before rising temperatures became a problem. Sound familiar? It was the same version of the salmon's life history that I had heard twenty years earlier on the Rogue River. This biologist could probably be excused for not being aware of his own agency's study done in the 1940s that suggested juvenile spring Chinook were rearing in the lower Willamette during the summer. However, I am confident he was aware of the life history studies from the Rogue and Sixes rivers. Those studies should have raised red flags that there was at least a strong possibility that juvenile spring Chinook salmon reared in the lower Willamette River in the summer.[33] As the biologist spoke, I imagined that I could hear the past smacking its lips as it devoured the present.

Fortunately, the City of Portland decided to fund a study to answer the question: Were juvenile spring Chinook rearing in the lower Willamette River in the summer or were they just migrating through in the spring? The Portland

Oregonian ran a major story on the study's findings and concluded: "After four years of study, researchers say they have overturned the conventional view of the lower river as a mere conduit for salmon racing to the ocean."[34] A summary of comments on the final report prepared by city staff included this conclusion: "The study has shown that in addition to serving as a migration corridor, there is active feeding and growth occurring as well. Rearing of juvenile salmonids in the lower river has been established. The lower river can no longer be viewed as a simple migratory corridor."[35] Fishery science once again showed that the salmon story and its simplifying assumptions did not reflect reality. The real question now is: How many times will fishery science have to dispel the same management myths? How long will the past continue to devour the present?

The extinction of salmon in over 40 percent of their historical range and the listing of salmon populations under the federal ESA stimulated a modest increase in the attention paid to habitat. However, even this slight divergence from the old salmon story produced a backlash from those who are intent on preserving the status quo. For example, in 2001[36] and 2003 the Oregon legislature tried to ensure that the failed salmon story continued to drive salmon management and recovery programs. Some politicians tried to make hatchery and wild fish equal by state statute. This would have meant that the recovery of listed stocks could be achieved by simply adding more factory-produced fish to Oregon's streams. Politicians were one of the beneficiaries of the salmon story and they wanted to maintain the political use of fish factories and avoid the inconvenience and cost of habitat protection and restoration. The statute, if the governor hadn't vetoed it, would have circumvented a legitimate scientific debate over the differences between hatchery and wild salmon and how these differences should be taken into account in management and recovery programs.

Here is another example. The Oregon Coast coho salmon were listed as threatened under the federal ESA in 1998. Following that listing, a group of citizens called the Alsea Valley Alliance filed suit in federal court to reverse the decision. The Alliance argued that National Oceanic and Atmospheric Administration (NOAA Fisheries, also known as National Marine Fisheries Service [NMFS]) should have included the number of hatchery fish returning to coastal streams when determining the coho salmon's status. Federal Judge Michael R. Hogan set aside the ESA listing, stating that NMFS had erred in its decision to exclude hatchery salmon. The goal of the lawsuit was clearly to prevent a change in the salmon story. There is no need to protect or prevent further destruction of salmon habitat as long as the rules governing ESA listings sanction the old practice of trading factory fish for habitat.[37]

In response to Judge Hogan's ruling, NMFS drafted a hatchery policy that defined how artificially propagated fish would be used to assess the status of salmon populations.[38] The Salmon Recovery Scientific Review Panel (SRSRP), a blue-ribbon panel of independent scientists, reviewed the policy. The SRSRP concluded that the policy did not reflect the published scientific research on the difference between artificially propagated and wild salmon and the implication of those differences for management and recovery programs.[39] According to panel members interviewed by the Union of Concerned Scientists, they were told to take those findings out of the report or "see their report end up in a drawer." Sometime later, the flawed policy was traced to a political appointee in the George W. Bush Administration. This same individual had previously advocated using hatchery salmon to boost the counts of endangered or threatened populations while he worked as a lawyer for the timber industry.[40] The past attempted to devour the latest science to protect the political value of the salmon story. Fortunately, the scientists on the SRSRP had the courage to confront and expose the myth.

In an effort to change the management of the Columbia River's water resources to achieve greater parity between fish and electric power production, the U. S. Congress enacted the Northwest Power Planning and Conservation Act (Power Act) of 1980. The Act created the Northwest Power and Conservation Council (Council), and directed it to develop a fish and wildlife restoration program. The Bonneville Power Administration (BPA) was to fund the restoration program through its power revenues.[41] The Council's initial Fish and Wildlife Program (FWP) was adopted in 1982, and it has been amended several times since then. The scope of the FWP and its cost constitute what may be the largest attempt at ecosystem restoration in the world.[42] The Council adopted an ecosystem approach to achieve its congressional mandate, which was consistent with the advice of fishery scientists,[43] and to the broader call for an ecosystem approach to the management of all natural resources.[44] The Council was attempting to craft a plan consistent with the latest science.

Twelve years after implementation of the first FWP, the Council recognized that its plan fell short of an ecosystem approach and that the "piecemeal efforts simply have not been effective." In 1994, the FWP underwent a major revision to bring it in line with the latest science. The entire plan was overhauled, but I'm going to focus on one section (Section 7—"Salmon Production and Habitat") and specifically the subsection, "Ensure Biodiversity." This subsection identified nine measures that the Council wanted salmon managers to implement with funding from BPA:

—Develop a policy to protect wild spawning populations.
—Evaluate salmon survival in the rivers and estuary to understand the ecology and capacity of the basin.
—Adjust hatchery releases to river carrying capacity.
—Collect baseline data on population status and life history of wild populations.
—Conserve genetic diversity.
—Review procedures for conducting population vulnerability analyses.
—Evaluate systemwide and cumulative impacts of existing and proposed artificial production projects.
—Establish a biodiversity institute.
—Reprogram existing hatchery stocks and facilities.[45]

These were key steps in the Council's plan to adopt an ecosystem approach to salmon recovery and to bring the FWP in line with the latest science. However, the Council has limited control over which parts of the FWP are actually implemented. The salmon managers in the Columbia Basin chose not to implement any of the measures in the biodiversity subsection of the new plan. Instead, they chose to implement a large number of hatchery projects. In its review of the suite of projects the fishery managers proposed to implement, the Independent Scientific Review Panel (ISRP) noted, "There is a noticeable discrepancy between the mix of projects actually funded and the ISRP's interpretation of the intent and priorities of the FWP."[46] The Council's attempt to implement a revised FWP that was consistent with the latest science was blocked by the salmon managers. They chose to remain within the constraints of the old salmon story and focus on fish factories. In making that choice, they acted as Fernand Braudel would have predicted.

Because the generation of hydroelectric power and the production of wild salmon both need water, there is a constant tension between the demands of the energy markets and the needs of the salmon. So far the salmon are losing the tug-of-war over the basin's water. For example, in 2001, BPA declared a flow emergency and abandoned the minimum flow safeguards for salmon in order to produce electricity. Even more troubling than the decision itself was the lack of any transparent criteria that would permit a review of BPA's decision and hold it accountable for taking such drastic action.[47] Did BPA abandon the concept of parity between salmon and power production or was this just another example of water management based on the mythological salmon, the socially constructed salmon engineered to survive in a river managed primarily as a money-making machine? BPA's decision, when viewed from the perspective of the salmon story and its factory-produced fish, should not come as a surprise. Nor should it be surprising that three years later BPA proposed decreasing spill in August so it could use the water to generate hydroelectric power and make an additional $47 million. BPA recognized that their proposed cuts in spill would kill juvenile salmon so they offered to use some of the extra millions to compensate for the loss.[48] Their proposal only made sense in the context of the industrialized salmon production system. It's OK to make a few bucks from the river machine and if it kills wild salmon, just replace them with factory fish. A federal court rejected BPA's proposal. The salmon's battle with the market-driven power system for a share of the water illustrates the point I made in Chapter 1: the return of the salmon is losing its role as the place-defining event because satisfying markets is given a higher priority.

The George W. Bush Administration's approach to the salmon's problem was entirely consistent with the salmon story and its reliance on fish factories. It cut protection for critical salmon habitat by 90 percent and agreed to count factory-produced salmon the same as wild fish.[49] Then they added an Alice in Wonderland dimension to salmon recovery. Dams on the Columbia and Sacramento rivers, they said, are part of the natural landscape and therefore could not be detrimental to salmon. Problem solved!

Dams and their problems for salmon are not new. They go back to the time before the arrival of Euro-Americans. Among the Indian myths is the story of how Coyote destroys a dam and frees the salmon so they can swim up river and feed the people.[50] The origins of that myth are lost in the fog of ancient time. It might be related to a large landslide in 1260 AD that blocked the Columbia River for a period of time.[51] Now compare the two stories regarding dams. In the Indian version, Coyote recognizes the problem posed by the obstruction and takes it down, freeing the salmon. In the Bush Administration's story, dams are considered a benign part of the landscape compatible with salmon. Can we really say that the modern myth is a better reflection of reality or that it is less naïve than the story of Coyote and the dam? Can we really say we have taken advantage of the power of Western science to craft a better story? Coyote, the trickster, is not the main character of today's salmon story; nevertheless, one gets the sense that there is a trickster wandering the corridors of the buildings where bureaucrats are busy spinning out solutions to the salmon's problem.

There was a time, early in my career, when anyone who talked about removing dams to help salmon was considered naïve or worse. Attitudes have changed to the point that dam removal to aid salmon recovery is not just talked about—it is actually being done. Recently, Gold Ray and Savage Rapids dams on the Rogue River in southwest Oregon were removed. Condit Dam on the White Salmon River is gone and so is Marmot Dam on the Sandy River, both tributaries to the Columbia River. Two large dams on the Elwha River in Washington State are being removed.

The removal of the two dams on the Elwha River is an example of how our salmon story persists even in the midst of seemingly positive and progressive change. The Elwha Dam removal is billed as an ecosystem restoration project, a very positive and progressive approach. To guide this ecosystem restoration, biologists from the cooperating state, tribal, and federal agencies prepared a restoration plan. The plan rightly states the importance of habitat: "Restoring and maintaining physical processes in the mainstem Elwha River habitat is the highest priority following dam removal."[52] Yet, in the 168 pages of the plan, only eight pages are devoted to the "highest priority" of habitat. Monitoring and evaluation are key tools in any recovery plan and are needed to assess the effectiveness of the plan's implementation and make timely course corrections where needed. Unfortunately there is not enough money in the Elwha restoration program to fund all the needed monitoring identified in the plan. OK, there is never enough money to do everything that should be done, but that being said, it was possible to scrape together $16 million for a brand new fish factory. Surprised? You shouldn't be if you have reached this point in the book.

Conservation organizations raised enough objections to the Elwha Recovery Plan that it was subjected to a review by a panel of biologists—the Hatchery Scientific Review Group (HSRG). I was a special consultant to the HSRG, in January 2012. The HSRG's report identified the positive and negative parts of the plan and made recommendations to improve it. In their report, in the section, "Likelihood of Success of the Plan," the HSRG said, "The greatest concern the

HSRG has about the likelihood of success of the restoration program is the use of an extensive hatchery program combined with the *lack* of a structured adaptive management process driven by an effective monitoring and evaluation program."[53] If the plan is revised as recommended by the HSRG, it will benefit the program. However, the hatchery is built and is operating. You can bet it will be used. Once again the past is devouring the present. The myth is given more validity than science.

∽〜

Fish factories and harvest management are the two most important elements in our salmon story. So far this discussion has focused on fish factories. While harvest and hatcheries are often considered separately as I am doing in this chapter, they are intimately linked. Biologist Phil Mundy summed up one particularly troubling aspect of the linkage between hatcheries and harvest: "The willingness to sacrifice vulnerable wild salmon stocks in order to harvest the bountiful hatchery returns of 2001 to 2003 and especially 2002 follows a long-established harvest management formula that has frequently led to disaster for conservation of wild salmon stocks in the Columbia River and elsewhere in the Pacific Northwest."[54]

Harvest is not just a word. It represents a concept that exerts a strong in-fluence on salmon management. Here is what my dictionary has to say about harvest.

Harvest—1 The gathering in of a crop. 2a The crop that ripens or is gathered in a season. b The time or season of such gathering. 3 The result or consequence of an activity.[55]

The word implies the gathering in of something that has been produced by human efforts. Harvest is the appropriate term when used to describe the gathering of crops produced by agriculture and aquaculture. Within aquaculture, the fish factory gave rise to the idea that we can "till the waters" and mimic the success of industrial agriculture.[56] Even though factory-produced salmon must spend a large part of their life outside human control in natural freshwater and marine ecosystems, there is the strong belief within salmon management that a crop is being produced and it must be harvested. Another consequence of the word harvest and its strong ties to agriculture is a focus on a single species—the crop to be harvested. This shifts attention away from the overall health of diverse aquatic ecosystems to the simple harvesting of a single species.[57] All of this is a predictable outcome of a salmon story that relies on an industrial production system. Single species management and its myopic focus on harvest have had an enormous impact on wild salmon. An incident that occurred during the preparation of a paper in 1988 illustrates the problem.

Willa Nehlsen and Jack Williams asked me to help them with a survey of the status of Pacific salmon stocks in the states of California, Oregon, Idaho, and Washington. I was in the midst of leaving my job at ODFW and moving to the Olympic Peninsula, so at first I hesitated to sign onto such a large project. However, Willa and Jack are two of my fisheries heroes, so I couldn't pass up the chance to work with them. As it turned out, the project, which was published in 1991 in *Fisheries Magazine* with the title, "Pacific Salmon at the Crossroads,"[58] was an important warning of the impending salmon crisis. While Willa, Jack, and I received a great deal of helpful cooperation from fisheries biologists in the region, there were salmon managers who neither saw a looming problem, nor saw any value in "airing dirty linen" in this paper. Most of those who disagreed with the paper's assertion of a widespread risk of extinction were involved in harvest regulation or hatchery management. They were advocates of the status quo, and the Crossroads paper challenged that status quo.

Willa sent a draft of the paper to salmon managers throughout the region for review and comment. A week or so later, I received a call from a manager working in Puget Sound. He told me that a few wild coho populations in his area were erroneously placed on our list of stocks at risk of extinction. I thought he meant that some of the coho populations that we said were at risk were actually healthy. I told him to send me the information he had on their condition and if we could agree that they were healthy we would certainly make the appropriate changes. He replied, "You don't understand, those wild stocks shouldn't be listed because their protection is not a high priority. If we have to protect those wild fish it will interfere with the harvest of hatchery-produced coho salmon in the same area."[59] The factory-produced salmon were harvested while they were mixed together with the wild salmon. If the weak wild stocks were protected, the hatchery stock would be underutilized. Protecting wild coho salmon introduced inefficiency into the industrial salmon production system; it interfered with the harvest of his crop of coho salmon. The salmon story and the need to maintain economic efficiency while supplying the market rendered those wild stocks irrelevant.

At the time of the phone call I was at the mid-point of my career and working on the Olympic Peninsula. For the previous twenty years I had worked on the effects of logging, dams, irrigation diversions, grazing, urban and industrial development, and the evaluation of hatchery practices—the tip of the iceberg to use John Livingston's metaphor. Personal experience like the incident described above and a growing interest in the history of my profession fomented a major change in my thinking about salmon management. Working on the Crossroads paper added to my growing frustration with the status quo and its failure to conserve wild salmon. I was beginning to realize that salmon management was part of the problem. The change in my thinking did not come all at once like a bolt of lightning. Rather, it grew over time as little drips and drops of experience coalesced into a different vision and a different interpretation of my accumulated knowledge. But a single triggering event can make you realize how much your view of the world has changed.

That phone call was one of those triggering events. After that conversation, I began to see the connection between the state of the salmon and a faulty management paradigm, and between the rapidly declining wild salmon populations and the industrial model of the salmon's production system. That salmon manager unwittingly exposed an inevitable consequence of the salmon story: wild salmon populations were irrelevant. They were more than irrelevant; they were impediments to the efficient harvest of the crop of engineered salmon. I had heard similar statements before, but didn't understand their significance and their deep roots in the flawed story. In retrospect, it should have been obvious that the close relationship between harvest and salmon factories left little room for wild salmon.

To explore the relationship between fish factories and harvest and its impact on wild salmon, I need to start with a little background information. In the protected environment of the hatchery, the survival of juvenile salmon from egg to smolt is higher than the survival of their wild cousins living in the river.[60] The hatchery eliminates things that kill wild salmon such as floods that scour the streambed, silt that suffocates the eggs, and several kinds of predators. The implications of this difference in survival are important. If the same number of eggs are deposited in a hatchery incubation tray and in a natural redd in the river, the eggs deposited in the hatchery will produce more smolts.[61] Assume that the goal of a fish factory is to match the number of smolts produced naturally in the river. Because of the survival difference, the hatchery requires fewer adult spawners (i.e., fewer eggs) to match the production of wild smolts. Since it needs fewer adult spawners, the hatchery population can theoretically sustain a higher harvest rate than wild populations. The belief that factory-produced fish can sustain a higher harvest rate is part of the dogma embedded in the salmon story and its industrial production system.

I am not convinced that hatchery stocks can indefinitely sustain high levels of selective harvest mortality. It is one of the unexamined myths, and it has been effective in selling the old salmon story. Fisheries are selective. They select the fastest growing or largest fish or they may select for the earlier or later migrating fish in a run. A high harvest rate in a selective fishery could alter the genetic structure of the population to its long-term detriment, especially if the selection targets a trait such as growth rate that is related to survival.

The bottom line is this: fisheries targeting mixed stocks of hatchery and wild salmon can have disastrous consequences for wild salmon. If the harvest of factory-produced fish is maximized according to the dogma, it will overharvest the wild population. Conversely, if harvest is regulated to protect wild populations, hatcheries will be flooded with more adult salmon than they need to meet production targets.[62] Harvest managers view the latter scenario as a "waste" of fish, and the avoidance of this waste is used to justify the overharvest of wild salmon.[63] In my earlier book, *Salmon Without Rivers*, I compared the attributes of the natural and industrial economies. In that analysis, I concluded that the idea of waste was derived strictly from the industrial economy. In the natural economy there is no waste—everything is recycled. Using the avoidance of waste to justify excessive harvest of wild populations is more evidence that salmon management is based on a simple, industrial model.

If egg to smolt survival is higher in the hatchery than in the river, why don't the hatcheries flood the rivers with so many salmon that there would be no need for harvest regulations, as was predicted by Spencer Baird, the first U. S. Fish Commissioner.[64] This hasn't happened, in part, because the survival of factory-produced fish, once they are released from the protected environment of the hatchery, is lower than that of their wild cousins, often quite a bit lower.

Hatcheries have never matched the natural productivity of rivers in their pristine state. Nevertheless, we continue to believe the myth, which sanctions the overharvest of wild salmon to prevent the waste of the industrial salmon crop. In some rivers, the spawning grounds are empty or nearly so where once thousands of wild salmon would have been busy creating a new generation. Overharvest isn't responsible for all the empty rivers, but it has certainly played an important role.

The term escapement is used to describe wild salmon that survive the downstream migration as smolts, the long ocean migration, and their return to the home stream to spawn; they escaped all the sources of mortality prior to spawning. The number of fish in the escapement is a measure of the efficacy of harvest management because it answers the question: Did the fishery harvest all the fish that it could have? The number of fish in the escapement, the distribution of those fish throughout the watershed, and their origins (hatchery or wild) provide input on several other important questions. Was the harvest adequately controlled so enough fish escaped to the spawning grounds to fully seed the habitat? Where a fishery targets a mixture of stocks, did enough fish from each stock escape the fishery? Was the escapement large enough to supply nutrients to the aquatic food webs and maintain carrying capacity of the habitat? Answers to those questions tell us if harvest management is supporting natural production and the health of the ecosystem.[65] However, when salmon management is based on a simple industrial model and the harvest of the factory-produced crop is the highest priority, many of these ecologically oriented questions lose their importance or relevance. As the story of the phone call and the quote from Phil Mundy indicate, the only question that appears to matter is: Did the fishery harvest the maximum number of factory-produced fish? Management consistently fails to achieve adequate escapements of wild salmon because it gives the harvest of factory-produced fish a higher priority than the health of wild salmon populations and their ecosystems.

In a 1985 paper, Michael Fraidenburg and Richard Lincoln, two biologists from the Washington Department of Fisheries (now the Washington Department of Fish and Wildlife), reviewed the management of wild Chinook salmon in the Pacific Northwest. Near the end of their paper they conclude: "Successful Chinook salmon management must embody the fundamental objective of providing enough fish for optimum spawning escapements." But they show evidence that: "Pacific coast river systems from northern California to southeast

Alaska are consistently 'under-escaped' by about 500,000 spawners per year ... Spawning escapements of some stocks are now more than 70 percent below optimum goals."[66] To illustrate the problem, they describe in detail the harvest and escapement management of the Klamath River Chinook salmon. It was in their words "a case of compromising standards."

In 1978, the Pacific Fisheries Management Council set an escapement goal for the Klamath Basin of 115,000 adult Chinook salmon. Two years later that goal was reduced to an interim target of 86,000 to prevent hardship on the troll fishery, with a commitment to return to the original target in four years. In 1983, fishermen were still experiencing economic hardship so the original escapement goal was put off for sixteen years and the escapement target reduced again, this time to 69,000 fish. But that figure included fish harvested in the river, so the actual number of salmon escaping the fishery to spawn was smaller, reaching a low of 31,500 fish in 1983. Managers then agreed to an escapement floor of 35,000 fish and that the actual escapement target would be set each year. The target for 1995 was 75,000 fish, which was still far below the original goal. In commenting on the Klamath River situation Fraidenburg and Lincoln said: "While this case provides an excellent example of how politics has influenced salmon management, it also illustrates how scientists and managers sometimes participate in regulating a fishery into over fishing."[67] They are right, but I would add that salmon managers are able to rationalize their politically motivated decisions because those decisions are consistent with the old salmon story and its industrial production system.

Fifteen years after Fraidenburg and Lincoln published their findings, Eric Knudsen surveyed the status of salmon escapements along the Pacific coast. He identified 1,025 management units composed entirely or mostly of wild salmon. They represented about 90 percent of the units in Alaska, Washington, Oregon, and California. Knudsen then rated the methods used to set the salmon escapement targets and the methods used to actually measure the escapement. The results were disappointing. Escapement targets for 80 percent of the management units identified in his study were set by methods rated as fair or poor or they have no targets at all.[68]

To add to this gloomy picture, Knudsen also found that the methods used to measure the escapement were questionable. For 93 percent of the salmon management units in Washington, Oregon, and California, managers collected escapement data by methods rated as fair or poor or the escapement was not monitored at all.[69] To preserve the wild salmon's life cycle and maintain the health of the ecosystem that sustains them, it is critical that managers set and achieve adequate escapement targets. However, this would require a new salmon story and management paradigm that is very different than the current one.

Overharvest, habitat destruction, dams, and hatcheries have done more than reduce the number of wild salmon. They have created a nutrient deficit of five to seven thousand tons of marine-derived nitrogen and phosphorous in the streams of the Pacific Northwest. This means the current salmon escapements are carrying only 5 to 7 percent of the nutrients of those large, historical runs. Streams that support coho salmon require 93 to 155 coho carcasses per kilometer to achieve maximum ecological benefit; however, the escapement target for Oregon's coastal streams in the 1990s was 26 coho per kilometer. The actual escapement between 1990 and 1995 was 2-7 fish per kilometer.[70] Nutrients carried in the bodies of adult salmon become part of the food webs—part of the salmon's habitat—and a critical part of the web of relationships that bind the salmon to their ecosystems.

Setting ecologically relevant escapement targets and adopting policies that ensure those targets are met would be a major step in the right direction. However, harvest management is an entrenched bureaucracy within fish and wildlife institutions and it has resisted change, especially change that would place the regulation of harvest in an ecological context.[71]

During my career, I served on several independent scientific review and advisory panels from British Columbia's Skeena River to California's Sacramento River. In aggregate those panels looked into all areas related to salmon recovery and management: forest practices, dam operations, hatcheries, predation by marine mammals and birds, water management, recovery plans, agriculture, and urban development. I am sure that no one enjoys having a panel of scientists looking over their shoulder, but most of the people involved in those reviews cooperated and willingly supplied information. They all cooperated except harvest managers.[72] Some harvest managers resented any review of their work. Most tried to respond to questions, but they found it difficult and frustrating to communicate with the panel members.[73] The panel members were usually scientists with strong ecological backgrounds. The two groups—harvest managers and panel members—viewed salmon through very different conceptual lenses. At times, I would step back and listen to the discussion. It was like watching individuals from different cultures with different languages trying to communicate.

In his book, *Sustainability*, Bryan Norton discussed the communication problems that develop between divisions or sections within the same institution. He coined the term "towering" to describe what happens when different groups within a single organization work within narrowly defined disciplines. They develop their own stories, vocabularies, models, and ways of viewing nature and in the process they lose the ability to effectively communicate with others outside their group. The barriers to communication also insulate the specialized groups from challenge and criticism.[74]

Towering and its effect on communication within the NMFS surfaced in a review of salmon harvest management conducted by the Salmon Recovery Science Review Panel (SRSRP), convened by NMFS in 2001. In the second paragraph of its report, the SRSRP stated, "Models used to set allowable harvest for fisheries are notoriously inaccessible and impenetrable for ecologists."[75] Donna Darm, Assistant Regional Administrator, Protected Resources Division of NMFS recognized that communication was impaired: "NMFS clearly did not communicate as well as we should have what our science is and what standard is underlying it. These are really smart people [the SRSRP]. Maybe we did not explain clearly enough."[76] It wasn't that the educational background of the SRSRP was lacking or that the harvest managers were more gifted modelers. The managers and ecologists had difficulty communicating because their respective conceptual models of the salmon production systems were so very different.[77]

This incident is an example of towering due to a growing divergence in the conceptual frameworks of fishery ecologists and managers. The communication problems created by the divergence are often attributed to a lack of skilled presentation rather than a fundamental difference in worldviews. A few scientists have actually analyzed the conceptual foundations of management programs and, as a result, realize that there is something more fundamental involved than poor presentation. Dan Bottom's landmark paper was the first comprehensive examination of fishery management's conceptual foundation and it opened the door to subsequent studies.[78]

One of those subsequent studies was triggered by the failure to halt the decline in salmon abundance in the Columbia River. Thirteen years after the Northwest Power Planning Council initiated its massive salmon recovery effort in the

Columbia River the abundance of salmon fell to a historic low of 749,000 fish (historical abundance was 10-16 million fish). Faced with these discouraging results, the Council, with some prompting from the Independent Science Group (ISG),[79] asked the ISG to undertake a review of the conceptual foundation of the salmon recovery program. I was a member of the ISG and contributed to the report. A published summary of the ISG's findings included the following statement:

> *After reviewing the science behind salmon restoration and the persistent trends of declining abundance of Columbia River salmon, we concluded that the FWP's [Fish and Wildlife Program] implied conceptual foundation did not reflect the latest scientific understanding of ecosystem science and salmonid restoration.[80]*

The ISG identified three key assumptions that characterized the flawed conceptual foundation:

> *The number of adult salmon and steelhead recruited is primarily a positive response to the number of smolts produced. This assumes that human-induced losses of production capacity can be mitigated by actions to increase the number of smolts that reach the ocean, for example, through barging, the use of passage technology at dams, and hatchery production.*
>
> *Salmon and steelhead production can be maintained or increased by focusing management primarily on in-basin components of the Columbia River. Estuary and ocean conditions are ignored because they are largely uncontrollable.*
>
> *Salmon species can effectively be managed independently of one another. Management actions designed to protect or restore one species or population will not compromise environmental attributes that form the basis for production by another species or population.[81]*

These assumptions are the foundation, the basic plot, of what I have been calling our salmon story or conceptual foundation. The first assumption says that habitat is irrelevant because technology can overcome any loss or degradation of natural reproduction. Fish factories or other forms of technology that boost the number of juvenile salmon migrating to sea will overcome habitat degradation and maintain the number of adult salmon. Technology is an acceptable substitute for healthy habitat and biodiversity.

The second assumption assumes the ocean is simply a vessel that fish factories can fill up. It ignores the way salmon life histories integrate the marine and freshwater environments. This assumption does not recognize that the loss of life history diversity due to the impact of fish factories and habitat degradation removes the buffer that diversity provides against fluctuations in the oceanic environment. The proximate cause of wild fluctuations in the abundance of salmon may be the changing ocean conditions over which we have little control, but those fluctuations originate in the loss of biodiversity occurring in fresh water, and that is a consequence of our salmon story.

The final assumption encourages single species management, which is consistent with the simplification and industrialization of the salmon's production system. It ignores the numerous relationships among species of salmon and between the salmon and their ecosystem.

The assumptions, beliefs, and myths that make up our salmon story are not unique to salmon fisheries management. Similar assumptions guide the management of other fisheries and the consequences are similar.[82] Atlantic cod in the waters off eastern Canada collapsed in 1992, with stunning social and economic consequences to coastal communities. That fishery had persisted for five hundred years and, in spite of the power of modern Western science, it collapsed.[83] There are signs that the cod are beginning to recover, but they are still at very low levels of abundance. Among the many studies of the causes and consequences of the collapse was one carried out by a sociologist, Alan Finlayson. He examined, among other things, the conceptual foundation that guided the assessment of the cod's status and the manager's harvest recommendations prior to the collapse of the fishery. Finlayson identified six assumptions that made up what I would call cod management's conceptual foundation. He called them a "techno–utopian" approach to cod management.

The universe is mechanistic and deterministic and its workings are governed by a few fundamental and unvarying laws.

The marine ecosystem and its sub-systems (in this case, commercially valuable fish stocks) are fundamentally robust. That is, they are relatively insensitive to small perturbations and tend to vary around natural dynamic equilibrium states.

These natural equilibrium states are dominated (or can be described and represented) by relatively few significant variables. In this case, they are fecundity, recruitment, natural mortality, and fishing mortality.

These variables are knowable and their effects on the stocks are simple, continuous, and can be realistically modeled by an equation with a small number of parameters. Therefore, they are predictable.

Science–based management can manipulate some of these variables (primarily fishing mortality). It can monitor the others to effectively control the system and produce (within certain broad limits) equilibrium states in general harmony with human needs and desires.

Having rebuilt the stocks to the desired level, they could then be maintained at that level by relatively minor adjustments in the total allowable catch. Long-sought-for stability could be brought to the fishery and its industries.[84]

There are strong parallels between the cod and salmon stories. Both simplify their respective production systems by defining them with a few controllable variables. Both narrow the focus to the number or biomass of target species, while ignoring the ecological relationships that produce them. Finlayson's "techno-utopian" term could be applied to salmon management's reliance on fish factories. Now there are calls for the release of factory-produced fish to rebuild the cod fishery.[85] Once again a complex ecosystem that sustained a valuable fishery is to be replaced by a simple industrial process. Although attempts to culture the cod go back a century, until very recently those attempts were unsuccessful. Will the new cod factories restore or replace the bountiful productivity of the wild cod stocks? Did salmon factories replace the productivity of the wild salmon runs?

In summing up this chapter's message, it's important to remember that Pacific salmon coevolved over millions of years with the riverine, estuarine, and marine ecosystems of the Pacific Northwest. Animals coevolve through the medium of ecological relationships. Because the salmon are a keystone animal they coevolved within a particularly dense web of relationships. Then humans entered the picture. Native Americans did little to disrupt the salmon's coevolved web of relationships; in fact, the Native Americans and their stories became part of the web of relationships. Then, a little over one hundred and sixty years ago, Euro-Americans flooded into the Pacific Northwest. The coevolved relationships that had sustained salmon for millions of years were thrown into chaos. The newcomers had no desire to fit into an ecological web; they wanted to simplify, industrialize, and control it. Perhaps this simplification was necessary as the only way that the managers could comprehend the nature of their job. It was just too difficult to comprehend and manage a keystone animal whose ecosystem stretched over a thousand miles or more. Simplification created a pliable and controllable resource and an ecosystem of a size and complexity that managers could wrap their minds around. The result was a transformation of the salmon sustaining-ecosystems, largely accomplished through the use of a powerful tool, the fish factory. Fish factories became the heart of a simple, understandable, pliable, and controllable salmon production system; a production system that could be superimposed onto rivers, replacing the complex and mysterious life history and ecology of wild salmon. A million years of evolution was turned upside down. The simple industrial production system conceived in the mind of humans was treated as the time-tested and proven way to make salmon. Wild salmon and the real ecosystems had to shed the lessons of their evolutionary history and adapt to the industrial production system or be extirpated.

So here we are today trying to recover wild salmon using a set of beliefs, myths, and assumptions that were originally intended to replace those wild fish. We look at the rivers, scratch our heads, and wonder why it isn't working. Throughout this chapter I have explained why I believe the current recovery efforts are not working. I have one more reason to add and it is this: numbers. The industrial production system stripped away all the messy ecological elements that salmon depended on for millions of years and shifted the focus to numbers of fish. So the salmon's problem is not defined in terms of the unraveling of the underlying ecological relationships, but simply in terms of the number of fish. Fishery managers haven't learned that declining numbers are the symptom and not the problem.

The focus on symptoms leads to the use of what biologist Gary Meffe calls halfway technologies—management and restoration measures that treat symptoms while ignoring the underlying cause.[86] When the salmon's problem is defined simply as low numbers, it inevitably leads to fish factories as a solution. In this regard, the salmon story is internally consistent, but consistently wrong.

Biologists managing a highly simplified salmon production system and focusing on symptoms instead of the real problems are often mystified by changes in abundance because the cause of those changes are "below the radar" of the salmon story. "It's perplexing. We don't have any answers," was Curt

Melcher's response to a major drop in the abundance of the Willamette spring Chinook run in 2006. Melcher was the assistant administrator of the ODFW's Fish Division. His attempt to explain the problem only exposed his confusion: "Remember that our forecast models are not biased. We can just as easily over predict as under. They have been so bad for the past two years that we might get one on the up side instead."[87] He was saying: We don't know why the system has failed, but maybe we will get lucky and the same mysterious reasons that cause declines might produce more salmon. The lesson that seems to be lost is this: you can simplify and industrialize salmon production and ignore the ecological processes that make up salmon-sustaining ecosystems, but eventually the things being ignored will return and extract a high price.

⌒⌒

Pacific salmon are a highly valued part of the Pacific Northwest, so highly valued that they have become a regional icon. In spite of that status, the salmon have been extirpated in over 40 percent of their historical range. In most of the watersheds where they still exist, they are in a sorry state. The listing of salmon throughout the Northwest under the federal ESA exposed the existence of a crisis—a crisis we created because our salmon story failed to conserve this animal that we value so highly. To protect and promote the recovery of listed salmon, all kinds of economic activities ranging from fishing, timber harvest, hydroelectric production, irrigation, and urban and industrial development have had to contend with an extra layer of restrictions. Councils, agencies, commissions, etc. struggle to resolve the crisis, but little or no attention is given to a revision of the salmon story that has served as a silent partner in creating and is now prolonging the crisis.

However, changing the salmon story is not simply an editorial challenge. The old story and the institutional structure that implements it coevolved over the last one hundred and thirty years. Salmon management and its underlying story evolved to fit into the institutional structure and vice versa. The institutional structure and the story are mutually supporting and so intertwined it is hard to understand the salmon's problem without examining both. The implications are important. If the story were rewritten so it reflected ecological reality, it could not be implemented within the old institutional structure. Likewise, a change in institutional structure would do little good if the new structure simply implements the old story. Both the story and the institutions are in need of change. So far I have told you about the story. In the next chapter, I take up the institutional part of the salmon's problem.

Side Channel 2
Thin Green Lines

Most people, he had said, go through life looking and never see a thing. Anything you see is interesting, from a chinch bug to a barnacle, if you just look at it and wonder about it a little. Then he would send me to the swamps or out in the boat or off along the beach with a firm command to look and tell him later what I saw. I saw plenty and in detail, whether it was ants working or a mink swimming or a tumblebug endlessly pushing its ball.

—Robert Ruark[1]

A stream meanders in the flat at the base of a low gradient slope. The course of the stream is clearly marked by a thin band of alders that snake across the clear-cut. It's a large cut extending down the slope to the stream and beyond, taking in most of the flat. I walked the length of the stream where it crosses the naked land, checking the strip of trees left in the riparian zone. The loggers followed the rules and left the required number of trees, but what they left is a puny substitute for the forest.

After checking the stream, I take a break before going on to the next clear-cut. Sitting on a large stump looking at the thin green line of alders, I think about the fate of the stream and its coho salmon. I think about the hubris that allows us to call this resource management.

I visited the site of this clear-cut several weeks before it was logged. At that time, the forest and its small stream were very different. The forest floor and the stream were in deep shade protected from direct exposure to the sun by a drippy green canopy. The shaded and moist understory was ideal habitat for the wood-sorrel, oxalis. The delicate herb with clover-like leaves covered the forest floor with a lush carpet of green. Today, the ground is dry and parched under the summer sun. The oxalis and their habitat are gone.

While I was thinking about the small stream and the uncertain future of its salmon, I notice a log about eighteen inches in diameter a few feet away. The sun is high—it is nearly noon—so the log casts a pencil-thin line of shade along its length. The line of shade is so thin I wouldn't have noticed it except for a single file row of little oxalis plants clinging to its protection. I glance around at other logs, but this one appears to be the only one of the right diameter and orientation to the sun to provide a razor's thin line of oxalis habitat. As I study these delicate little plants it reminds me how tenaciously life in all its forms can cling to even the smallest remnant of habitat. I imagine I can hear their fragile cries for help, in what to them must have been the equivalent of a raging inferno at the edge of their precarious refuge.

A question pops into my head while I am looking at the row of oxalis plants. What if we could see streams and stream habitat the same way a juvenile trout or salmon experiences it? What if we could see limiting factors in salmon habitat as clearly as I could see those oxalis plants limited to their pencil-thin line of shade? Of course we can't see salmon habitat from the salmon's perspective with perfect clarity. The aquatic medium is foreign to us and to our sensory attributes, but it is not the only impediment to clarity. Our cultural filters also influence what we see. For some, those filters can so reduce their vision that small streams and the wild salmon in them simply disappear. All they see is fish in the boat—hatchery fish in the boat.

If we could see salmon habitat with perfect clarity, in many streams the vision would look like this single row of oxalis desperately clinging to a pencil-thin line of shade. A fisheries professor once told me that only a salmon really knows what constitutes salmon habitat. He went on to say that our best efforts to understand salmon habitat will always be imperfect. However, the degree of imperfection will vary depending on the degree to which we allow our cultural filters to distort what we see. It's hard to pay attention to the details of the salmon's world, if our vision of that world is distorted by an industrial model that renders those details irrelevant. Such thinking has led me to this conclusion: we will have difficulty in restoring even a small part of the salmon's productivity until we learn to pay attention, until we learn to see the details as clearly as our biological sensory attributes will allow. An important part of the difference between Native Americans and Euro-Americans and the reason why the Indians were able to coevolve a sustainable relationship with the salmon was this: they paid attention. They paid attention to the details of the world they lived in.

We have a choice. We can continue to treat the natural world as though it were a facsimile of the industrial economy where ecosystems are machines and the resource manager's job is to pull the levers and push the buttons that control them. Or we can pay attention to nature and see it on its own terms, as best we can, and in the process deepen our sense of place. I know I will never be able to see the ecosystems I inhabit with full clarity, but it seems to me to keep trying is a nice way to spend the span of time allotted to me.

Before I moved from our home in the Olympic Mountains near Sequim, Washington, I took a walk around the yard to say goodbye to several friends. One of my stops was at a spot near the back of the house that was always in the shade. There on a small patch of shaded ground was a thick, green carpet of oxalis. I said goodbye and left them there. They were looking a lot happier than when we first met.

Chapter 3—The Meeting

To whom can responsibility ... be attributed? Or do we live in a context of 'organized irresponsibility'?

—Urlich Beck[1]

We have met the enemy, and he is us.

—Walt Kelly's Pogo Earth Day Poster, 1970

Four long tables are arranged in a square at one end of the large room. A tangle of wires and electronic devices fills the open space in the middle of the square. At three- to four-foot intervals, the small round heads of microphones rise from the tables on long skinny necks like a line of mesmerized cobras. The men and women who sit at the tables are the administrators of agencies and institutions that manage salmon in the Columbia Basin. To a large degree, the fate of the salmon is in their hands. They talk in a slow, measured cadence as though each word, idea, or sentence has to be inspected, and its implications carefully examined, before being set loose in the room.

The location is not important; it could be anywhere in the Pacific Northwest. In fact, meetings like this are held throughout the year in Sacramento, San Francisco, Boise, Salem, Portland, Olympia, and Seattle. Hundreds, maybe thousands of meetings throughout the year, and all of them are trying to solve the same problem. They are searching for a sustainable relationship between salmon and people in the Pacific Northwest. This particular meeting is in

Portland, Oregon, and it concerns the salmon in the Columbia River, although the meeting's purpose would be hard for an outsider to determine. I've been listening for forty-five minutes, and the word salmon is not among those being carefully released into the room. The talk is about budgets, authorities, future funding, and who should have a seat at the table. Those subjects are important to the institutions these men and women represent and they must believe they are also important to the salmon, but I have my doubts. The agenda shows topics that deal directly with salmon so I decide to stay.

While the meeting plods through bureaucratic trivia, I think about the salmon's problem, not the well-known issues at the tip of John Livingston's iceberg, but the problem being demonstrated here in this meeting. These administrators, all of whom have a genuine concern for the salmon, are not aware that they are part of the problem. The problem I have in mind is the very institutional structure that these people represent—the institutional structure that was supposed to protect and conserve the salmon and their habitats. In spite of the huge expenditures on salmon restoration, this problem receives no serious attention. And that's understandable because if the institutional problem were given serious attention, these folks would have to conclude along with Pogo that "we have met the enemy and he is us."

Over a hundred years ago President Theodore Roosevelt called attention to the problem of salmon management's institutional structure. In his State of the Union Address for 1908, Roosevelt said that the dysfunctional institutional structure could push the salmon in the Columbia River and Puget Sound to extinction. He was referring to the management of salmon harvest, which was independently regulated by the states of Oregon and Washington in the Columbia and by Washington State and Canada for the stocks in the Puget Sound area. The management agencies could not agree on a coherent set of harvest regulations for salmon in the lower Columbia River or for salmon crossing the international boundary in Puget Sound. The result was chaos in the fisheries and persistent overharvest. President Roosevelt threatened to federalize salmon management, if the problem was not resolved.[2] Roosevelt's threat led to the Columbia River Compact, which sets uniform harvest regulations for the lower Columbia River. Several years later the International Pacific Salmon Fisheries Commission was created to manage the international fishery on Fraser River salmon. While these were positive steps, they did not recognize nor deal with the full scope of the problem.

In the late 1930s and early 1940s, the states of Oregon and Washington concluded that a workable solution to the decline of salmon in the Columbia River required a change in salmon management's institutional structure. A special subcommittee of the Washington State Senate concluded that salmon management was fragmented among too many institutions to be effective. The senators emphasized that "We are hopelessly defeated in obtaining any solution to the Columbia River fisheries unless we simplify our administration over this resource."[3] The Oregon State Planning Board reached a similar conclusion.[4] Sixty years later, the National Research Council concluded: "The current set of institutional arrangements is not appropriate to the bioregional requirements of salmon and their ecosystems," and further that "the current set of institutional arrangements contributes to the decline of salmon and cannot halt the decline."[5] Over a hundred years ago President Roosevelt told us about the institutional problem and, since his warning, we have been reminded of its continuing existence, but we have done nothing about it, except watch it grow bigger.

Salmon thoroughly penetrate the ecosystems of the Pacific Northwest. Their long migrations take them across the jurisdictions of several federal, tribal, state, city, county, and private institutions. All of these organizations make decisions that affect the salmon's ecosystem and some of them have primary missions that conflict with salmon conservation. For example, the primary missions of the state and federal forestry agencies or of the public and private institutions that operate hydroelectric facilities can and often do conflict with salmon conservation. From the institutional perspective we cannot see a coherent management structure that covers the salmon's entire ecosystem. What we see is an ecosystem fractured into pieces more like what one sees when looking into a kaleidoscope.

An ecosystem fragmented among several institutions is a maze of bureaucratic boundaries that are vigorously defended. Crossing one of those boundaries and poking your nose into another agency's territory can be detrimental to a biologist's career, and an agency that strays over that boundary could find its budget cut. During the seventeen years that I worked for the Oregon Department of Fish and Wildlife (ODFW), I recall being told that there was little we could do to protect salmon habitat because it fell in the domains of other state or federal agencies such as the Department of Environmental Quality, the Environmental Protection Agency, state and federal forest management agencies, the Department of Water Resources, the Corps of Engineers, BPA and so on.[6] To simply ignore habitat fails the common sense test. The fragmented institutional structure and its bureaucratic boundaries create a plausible excuse: it makes habitat someone else's responsibility.

The people of the Pacific Northwest created institutions whose mission was to manage the Pacific salmon for the benefit of present and future generations. Why would biologists and administrators maintain an institutional structure that impedes their mission? Why wouldn't they be demanding change? I'll offer two possible explanations. The institutional structure and salmon management's industrial production system have been coevolving for more than a century. As they coevolved, each adapted to and reinforced the other. The result is a management paradigm that gives primary importance to hatcheries and harvest regulation because those two activities fall wholly within the boundaries of the ecosystem fragment allotted to fisheries. It defines the agencies' comfort zone, which is why, in spite of repeated warnings of its damaging effect on wild salmon, there has been no serious attempt to change the fragmented institutional structure. Another reason for the persistence of the fragmented institutional structure is that the people and their political representatives have been told over and over that salmon recovery is on the way. New programs are unveiled with the promise of doubling or tripling the runs of salmon. The runs continue to decline, generating more programs and promises. Unfortunately, no one with authority to demand accountability asks: What happened to the previous promises? The experts and administrators are trapped in an institutional structure and a set of assumptions and myths (the story) that cannot recover the salmon and indeed are contributing to their decline. It is a trap they themselves and their predecessors helped construct.

The salmon story and the institutional structure coevolved in a self-reinforcing relationship that has a set of unwritten rules defining rational behavior. Individuals working within this system may make a decision that to them seems highly rational, but the same decision, viewed from outside the system, lacks common sense. I witnessed this ability to rationalize detrimental behavior one day while I was in Salem to attend a hearing on the fish and wildlife budget. Later that day, I had some time to kill before catching a ride back to Portland, so I sat in on a meeting between an administrator at ODFW and a group of environmentalists. They were discussing language changes in Oregon's proposed Threatened and Endangered Species legislation. The administrator was proposing language that actually weakened the bill and it wasn't sound biology. I asked him about it on the ride back to Portland and he told me that ODFW had to be careful. If we supported strong endangered species legislation, others such as Oregon Department of Forestry would view it as a threat.

Why would this fish and wildlife administrator not side with the environmentalists and strengthen protection of threatened or endangered species? Why would he worry more about how the Department of Forestry viewed the legislation than the environmentalists? The environmentalists should have been one of ODFW's natural allies. Decisions like the one just described are often attributed to incompetence, weakness, or devious political maneuvering, but those terms are not adequate explanations. After I started research on the history of salmon management, I realized what the administrator was really telling me that day on the ride back to Portland. Staying inside our little fragment of the ecosystem and not poking our noses into another agency's fragment trumps the protection of fish and wildlife.

Science tells us that we should be taking an ecosystem approach to the management and recovery of Pacific salmon, but such an approach is not compatible with the current fragmented institutional structure. There have been attempts to overcome this impediment, but the bureaucratic boundaries and the special interests those boundaries protect are a formidable obstacle. So the coevolved institutional structure and the industrial production system remain even though they clash with the salmon's unique life history and its extended ecosystem.

In Chapter 2, I introduced W. F. Thompson's concept that the salmon's life history is a chain of connected habitats that they must pass through at the appropriate time. Each link in the chain of habitats supplies the resources and environmental attributes needed to complete the salmon's life cycle.[7] The chain of habitats stretches for thousands of miles from headwater spawning areas out to the north Pacific Ocean. A break in the chain (degraded habitat) can diminish or eliminate the population of salmon following it. Recently, Canadian biologists added some clarity to Thompson's "chain of habitats" concept with the following:

[E]ach species has a linked series of discrete life history stages that are tied to a particular habitat or part of it, and there can be a range of ages over which the various stages occur. However, within any population, each stage has to occur within a fixed seasonal time range. Therefore, for each species and population within it, life history is carried out through a series of space- and time-dependent stages.[8]

Those same biologists went on to say that a four-year-old sockeye salmon from the Fraser River must pass through twelve different habitats or twelve links in Thompson's life history-habitat chain.

Thompson's use of a chain as a metaphor for the salmon's life history helps illustrate one of the reasons why the people in the Pacific Northwest have spent so much money on salmon recovery and achieved so little. It also helps illustrate the clash between the salmon's life history and the fragmented institutional structure. Consider a salmon population following a life history pathway with twelve habitat links in the chain. Now let's say four of the links are broken—access to a critical habitat is impaired or the habitat is degraded to the point that salmon cannot make use of it. Through heroic efforts, three of those links are repaired. Those heroic efforts will not yield the expected increase in salmon because the chain is still broken. Recall the saying, "a chain is as strong as its weakest link."

Toward the end of my career, I served on several independent scientific review panels. This experience gave me the opportunity to review dozens of salmon restoration projects. Through that whole experience, I cannot recall a single instance when a project leader said his or her program was a failure. However, after reviewing one "success" after another for all those years, it is not possible to step back and see that they were making a real difference in the status of the salmon. Why the apparent disconnect between local successes and the poor response at larger landscape scales? There are a lot of hard-working individuals fixing links in the habitat-life history chain, but as far as I can tell, few if any of them are stepping back and looking at those individual projects from a whole ecosystem or whole life history perspective. No one is making sure all the links in the chain are in good shape. The narrow focus on individual links rather than the whole chain is reinforced by the way ecosystems are fragmented into narrowly defined institutional jurisdictions.

When Theodore Roosevelt expressed his concern over fragmented salmon management in 1908, he was referring to the inability of two state agencies (from Oregon and Washington) to enact consistent harvest regulations in the Columbia River. While President Roosevelt correctly identified a budding problem, he probably didn't foresee how much it would grow. For example, a juvenile salmon that leaves the Lochsa River in Idaho this spring will pass through the jurisdiction of seventeen different salmon management agencies[9] and many more public and private agencies and institutions whose decisions can affect salmon habitat. The problem that President Roosevelt identified more than a hundred years ago has evolved into a convoluted institutional morass that the Snake River Salmon Recovery Team called "jurisdictional chaos."[10] Fragmented responsibility for the salmon's ecosystem not only makes the salmon's long life history pathways vulnerable to disruption, it is extremely difficult to mount an integrated, whole life history or whole ecosystem restoration program. A recent event illustrates this problem.

After $53 million had been spent to develop an ecosystem approach to the management of seventy-one million acres of federal land in the interior Columbia Basin, the program fell apart in the early stages of implementation. The scientific assessments and the resulting databases that were part of the planning process will be valuable tools for land and water managers, but a fragmented institutional structure and conflicts among special interest groups prevented the parties from reaching a binding agreement on how the program as a whole should be implemented.[11] According to an article in the Portland *Oregonian* the failure of the interior Columbia Basin plan may have killed an ecosystem approach to the management of federal lands.[12]

The federal agencies involved are independently using the scientific assessments to implement parts of the plan, but the inability of the various parties to the plan to reach agreement on its implementation is a bleak sign for the future of salmon recovery. Long migrating species like Pacific salmon require whole ecosystem and whole life history approaches to restoration, but humans constrained by the current salmon story and a fragmented ecosystem have not been able to imagine and/or implement an institutional structure capable of managing salmon at the scale of their extended ecosystem.

I think about all this as the people at the table pick through bureaucratic trivia. As I listen to the administrators acting out their tragic roles in the "jurisdictional chaos," my mind wanders again back to the 1940s and the attempts by the Washington State Senate and the Oregon Planning Board to improve salmon management's institutional structure. It's a major shortcoming, a tragedy really, that none of the people in this meeting are probably aware of those efforts. Even if they were aware, I'm certain that none of them would have the courage of B. M. Brennan. He was the director of Washington State's Department of Fisheries and he was willing to sacrifice part of his institution's power and authority in order to create a management structure that benefited salmon.[13] He was willing to relax the bureaucratic boundary of his fragment of the Columbia River ecosystem to improve the management of Pacific salmon. So far in this meeting I have seen no evidence that the fate of the salmon is a higher priority than the maintenance of bureaucratic territories.

The meeting drones on with the participants still bickering about funding and budgets. The people in the audience are getting restless, shifting in their uncomfortable folding chairs. A man sitting in front of me closes his notebook and puts his pen back in his pocket. He is looking out the window at pigeons on the roof of the building across the street. Like me, the others in the audience are here to listen to a discussion of things important to the salmon. So far, all we have heard are things important to agencies and bureaucrats.

The listings of Pacific salmon under the federal ESA and the ensuing possibility of major disruptions in parts of the region's economy have made these meetings of interest to a broad range of folks, especially those with a financial stake in salmon recovery. To avoid or reduce the potential economic impact of the listings, citizens of the Pacific Northwest now invest nearly a billion dollars a year on salmon recovery.[14] However, the injection of that much money into salmon management's highly fragmented "jurisdictional chaos" has a serious unintended side effect. Fish management agencies competing for a slice of the funding shift their emphasis from the common goal of salmon recovery to the goal of maximizing their share of the recovery dollars. Sociologist Rik Scarce attributes the following to someone participating in the allocation of salmon recovery funds: "[F]unding decisions are highly political, part of a 'good old boys' network that primarily benefits those who are well connected."[15] Scarce's book clearly describes what I observed at the meeting: "good old boys" wrangling over their budgets. The condition described above creates an atmosphere that is poisonous to the acquisition, incorporation, and implementation of the best science in salmon management and recovery programs. It leads to stifled science, which is another subject that receives little public discussion. To its credit, the Snake River Salmon Recovery Team did discuss the problem in its 1994 plan.[16] Stifled science has been around for a long time. For example, it surfaced fifty years ago at a conference organized by the governors of the Pacific Northwest states and Alaska to discuss mutual concerns regarding the status of Pacific salmon.

Here is how W. F. Thompson described stifled science in a published commentary on the governors' conference:

> It was disquieting that there was little argument or discussion among the scientists present, as there might have been had they been free of controls. Each spoke as a representative in one way, or the other, of his organization, as though departmental 'policies' were involved in anything they might say. No antagonism or differences of opinion appeared even if present. This is not a healthy or normal state as far as scientists are concerned, because it is in diversity and originality of ideas that there exists opportunity for improvement or change, so badly needed in fisheries biology. It was most apparent that organizational controls dominated.[17]

Stifled science leads to a bureaucratic procedure called consensus management. Under the rules of consensus management, each institution wields considerable power because it only takes a single veto to derail a proposed course of action. Consensus management gives extraordinary power to the least informed and most politically motivated—to the "good old boys" mentioned earlier. Consensus management intertwines science, policy, and bureaucratic politics so tightly that those who become enmeshed in it find it hard to tell them apart. The Snake River Salmon Recovery Team called this management system "jurisdictional chaos, no one in charge, important decisions not based on science, and stifled science."[18]

Back at the meeting, a series of heated exchanges disrupts the discussion of bureaucratic trivia. One individual challenges the right of another to be at the table discussing important salmon issues. The accuser conveniently ignores the fact that up until now the word salmon has yet to be uttered and the real issues facing salmon have yet to be discussed. The challenger has so tightly conflated bureaucratic issues with the salmon's real problems that to him they are one and the same. He is not alone. In my travels through the institutional labyrinth dealing with salmon management and recovery, I have, from time to time, glimpsed the surreal condition where survival of the salmon institution or its policies is given higher priority than the survival of wild salmon.

This meeting, like many others, should have an empty chair at the table to symbolize that the wild salmon or any real advocate for them is notably absent. I play with the thought of a salmon sitting at the table. What would it say? What opinion would it have of salmon management? What kind of report would it give to the salmon king after the meeting? Would the Salmon King ever again send his subjects up the rivers to feed humans? If a salmon did come to this meeting to observe those entrusted with her fate, she would probably leave shaking her fins in disgust.

The brief dispute is over and the meeting resumes its mind-numbing discussions. To escape for a few minutes, my mind moves beyond the business suits, conference tables, and the cautious words about budgets. I leave the city and its concrete canyons, the factories, and suburbs, and go deep into the real canyons where the great rivers of the Northwest divide into thousands of small streams.

In one of those streams eight hundred miles from the ocean, water flows over a log and plunges into a pool. Small Chinook salmon about four inches long hide in the shadow of an undercut bank at the edge of the pool. They occasionally dart into the open after an insect. Their behavior appears normal, but under their scales big physiological changes are taking place to prepare the small salmon for the long migration down the river to the sea. Under natural, pristine river conditions, the juvenile salmon's migration to sea would be fraught with danger. Today, in the "developed" river, the migration is more dangerous than ever.

Dams with their slack water reservoirs full of predators are a threat to the migration of both adult and juvenile salmon. The nursery streams, where the juvenile salmon may spend a few months or a year or more, are dewatered by irrigation withdrawals or heated to a lethal temperature because cows or loggers have stripped away the riparian trees and shrubs. Unfortunately for the salmon, they do not escape the perils of development when they leave the river and enter the ocean. In fact, they face a new and growing threat. Climate change is degrading critical biophysical processes in the ocean with serious consequences to marine ecosystems.[19]

I keep wondering: How is this meeting solving the problems facing the little salmon in that pool so far from the ocean? In the meeting, the slow, measured words are carefully piled on the table. I am getting frustrated with the disconnect between what I'm hearing and what needs to be done to protect and recover wild salmon. The leaders of the agencies represented here are supposed to fight for the salmon. Do they really believe that what they are doing here is solving the salmon's problems? Why doesn't one of the administrators stand up and ask the question: How are we doing? How has this massive recovery effort materially improved the well being of wild salmon and their ecosystems? And then demand an answer. Not only demand an answer, but the evidence to back it up. I have never heard that simple question asked: How are we doing? How are we really doing?

Just before lunch, the word salmon finally makes an appearance at the meeting. A biologist makes a presentation to the group on flow augmentation—water released from storage reservoirs to flush groups of juvenile salmon through the Columbia River's mainstem-reservoirs. I smile on hearing the term, because it reinforces the view that the ecosystem is a machine and the river a simple conduit to the sea. To operate this machine, pull a lever or turn a valve, and send a pulse of water along with juvenile salmon down the conduit and out to sea. Running the river like a machine and the heavy reliance on fish factories are the two sides of the same coin.

Migration through a conduit allows no ecological connections between the river and the salmon. It denies the salmon's need to pass through and maintain a relationship with the chain of habitats where their life histories are played out. Thinking of the river as a conduit reduces the threat to the Lords of Yesterday, the economic powers that benefit from maintaining the ecosystem in its fragmented condition.[20] In the Columbia River water is money. It is money that the Lords of Yesterday reluctantly concede to the salmon. The augmented flows are one of those concessions, but it's a concession bound by rigid rules about what can and cannot be done. It's another small fragment partitioned from the larger ecosystem by strong bureaucratic boundaries. Those boundaries were imposed on the program to ensure that the river remains a simple conduit to the sea that will pose no threat that the salmon's ecological connection to the river will be reestablished or that the system of salmon hatcheries will be compromised.

The biologist presents data showing that the pulses of extra water appear to move the juvenile salmon downstream at a faster rate. At the end of the presentation, one of the administrators at the table asks the obvious question, "Does this mean that the flushed juveniles return as adults in greater numbers?" The biologist gives the questioner an evasive nonanswer. The questioner is from a federal agency and is new to the Northwest. He hasn't been exposed to the myopia induced, encouraged, and condoned by the fragmented institutional structure. The person asking the question unknowingly put the biologist in a difficult position. Augmented flows are a touchy subject and going beyond the narrow mandate of the program has caused political backlash and attempts to eliminate the entire program.[21] As I mentioned earlier, a biologist can get into serious trouble by crossing one of the bureaucratic boundaries.

The man sitting in front of me pulls out his pen and notebook, but as the discussion runs its course to a dead end, he puts the pen and notebook away and once again focuses his attention on the pigeons across the street.

By this time, the men and women around the tables have produced thousands of words, and I'm sure they feel that they have made progress in resolving the salmon's problem. They don't realize that they are prisoners of false optimism derived from a flawed story. The numerical models, the sophisticated technology, the immense size of the effort give the illusion that the region is really doing something to save the salmon. But strip away the new technology and at the core of the recovery programs are the same old myths, the same old story. These administrators have huge budgets and the latest technology at their disposal, but they are trapped in the old story while, unwittingly, they lavishly implement the status quo.

When the meeting ends, there are pats on the back, handshakes, and smiles all around. My thoughts drift back to that small stream in Idaho and the little silver fish that is about to start a long and dangerous journey. I know what lies ahead of it, and this meeting has done nothing to ease my fears.

An institutional structure that fragments ecosystems into pieces dominated by economic or bureaucratic Lords of Yesterday is an important part of the salmon's problem and a major impediment to recovery. But this does not complete the institutional story. The internal structure of salmon management agencies evolved to comply with and function within the bureaucratic constraints of fragmented ecosystems. So even if the jurisdictional chaos of the fragmented ecosystems were corrected, the full benefit would not be realized until there is a corresponding change in the internal structure of the management agencies.

To get a clear picture of this aspect of the salmon's problem, travel back to the later decades of the nineteenth century when state and federal governments established agencies to manage and conserve Pacific salmon. Because the salmon fishery and the canning industry were rapidly expanding, the recently appointed salmon managers decided their highest priority was to regulate and maintain order in the fishery. In addition, they firmly believed fish factories would add to the abundant wild runs and enhance the profits of the industry. This led to the management program organized around harvest regulation and hatcheries that I described in Chapter 2.

Harvest and hatchery management are still the strongest divisions within fish and wildlife agencies and they still exert the strongest influence on policy formulation. But today, rivers have been developed, habitat degraded, salmon populations depleted. The problems salmon managers face in the twenty-first century are very different than those that existed in the late nineteenth century. We cannot continue to harvest with little regard for ecologically adequate escapements, nor can hatchery ponds be used in lieu of habitat for ESA-listed populations. Artificial propagation and harvest management will continue to be important elements of salmon management, but that role must be supportive of and subordinate to the goal of increasing the health of salmon-sustaining ecosystems and the survival of both natural and artificially produced salmon.

Leaders of salmon management institutions constrained to operate within the bureaucratic boundaries of fragmented ecosystems and locked into the myths and assumptions of the old salmon story have recently been confronted by a growing number of organizations (such as watershed councils and environmental and native fish organizations) and the general public who are increasingly skeptical of the status quo. Administrators try to convince the public that their agency is responding in constructive ways to new information by becoming skilled in adopting the appropriate vocabulary. Words like ecosystem, biodiversity, wild salmon, natural production, recovery, habitat, and resilience are sprinkled throughout their reports and public statements. The use of scientifically appropriate vocabulary to draw attention away from outdated assumptions and myths corrupts language, impedes effective communication, and replaces thoughtful analysis of the salmon's problem with superficial musings. Using the correct vocabulary is not a substitute for a change in what an agency actually does.

Recall the salmon manager in Chapter 2 who was willing to sacrifice wild coho salmon stocks in order to maximize the harvest of factory-produced fish. Here is a postscript to that incident to illustrate the facile use of appropriate words. A couple of months after the paper "Pacific Salmon at the Crossroads" was published, it generated a great deal of interest by the media. The fate of wild salmon suddenly became an important issue. One afternoon I was working in my office. The radio was tuned to the local public station. I was fully absorbed in my work, paying little attention to the radio. Then the words *threatened wild salmon* caught my attention. The station was hosting a panel discussion on the status of wild salmon. Among the panel members was the supervisor of the individual who showed such little concern for the wild coho salmon. According to this administrator his organization had always given highest priority to the conservation of wild salmon. He used all the right words and sounded very sincere. I'm sure most of the people listening believed what he said.

Honest attempts to break with the status quo are difficult to implement because the individuals promoting the change fail to appreciate how resistant to change an agency can be. They don't understand the way fragmented management of ecosystems and the internal power structure of fish and wildlife agencies reinforce the status quo. During my career I have seen the best intentions of individuals trying to change the status quo thwarted again and again. Here is one example.

In 1978, a progressive Oregon Fish and Wildlife Commission tried to make a fundamental break with the status quo when it adopted a policy to protect and conserve wild fish. It was a contentious issue and the subject of much discussion and debate within ODFW. The policy finally adopted was a major departure from the status quo and if fully implemented it could have been the first step in revising the salmon story.

Five years after the Wild Fish Policy's adoption, I was appointed assistant chief of fisheries at ODFW and, since I had a keen interest in the Wild Fish Policy, I wanted to see how it was being implemented by the managers in the field. I sent a questionnaire to the district biologists and talked with many of them. To say that I was disappointed in the results of my survey would be an understatement. Although the managers were generally aware of the policy, they were not familiar with its actual language and how it should be applied to their day-to-day work. It was clear to me that some had not actually read the policy and others had totally misinterpreted its purpose. I learned that implementation was falling far short of the commission's expectation.

I now know that expecting the managers to fully implement the Wild Fish Policy was, in fact, giving them a difficult and maybe impossible task. They were being asked to implement a major change, a change that was antithetical to the old salmon story and its industrial production system. The managers were expected to reconcile and use two, often incompatible, approaches to their management practices—a strong wild salmon conservation ethic and an industrial production system—without having the tools and the organizational support needed to achieve that reconciliation. The new policy was inserted into an organizational structure designed to implement the old story. It was an orphan wondering the halls of the bureaucracy with no place it could call home. New policies like the Wild Fish Policy that are superimposed on the old story and organizational structure result at best in superficial changes. The words may change, but the actions are the same. They are similar to the changes made in the false fronts of movie towns. One day the set may be a quiet town in middle America, the next day it is a wild town on the edge of the frontier. Seen through the lens of the camera, the towns are very different, but behind the false fronts everything remains the same. If you want to change the flow of priorities in an institution and ensure the implementation of new policies, it's important to get inside the organization and change the basic plumbing. That is rarely done.

Over a hundred and thirty years ago, Spencer Baird, the first U. S. Fish Commissioner, quite accurately described the three main threats to the persistence of the salmon: overharvest, habitat degradation, and dams.[22] Yet, in spite of this knowledge, our attempts to "manage" salmon have failed because they were based on a flawed model of the salmon's production system, which was also introduced by Baird. That model was derived from the belief that humans must control nature and use technology to circumvent ecological relationships and processes. For more than a hundred and thirty years, the model and the institutions that implement it coevolved into a tightly bound relationship that persists in spite of the accumulating scientific evidence that a different approach is needed.

It is not a new problem that the men and women were discussing in the meeting described earlier in this chapter or in any of the thousands of similar meetings that take place every year. Fundamentally, it is a problem that humans in the Northwest have struggled with for at least four thousand years. Both Native Americans and Euro-Americans have had to solve the same management problems within the constraints of their cultures.[23] The people in this meeting were not thinking about their place in the four-thousand-year struggle to achieve a sustainable relationship with salmon, because their approach to this problem is devoid of a historical context. Their simplified model of the salmon's production system and its machine-like operation renders history irrelevant. They work only in the present, disconnected from the historical roots of the problem they grapple with. Released from the constraints of history, they will repeat the mistakes of the past, so despite their good intentions and new scientific information, in the end, they will get the same results. Historian Robert Bunting said, "Unless the people of the Pacific Northwest envision a historically informed future, they face the prospects of losing the environment that has so fundamentally shaped what it has meant to be a Pacific Northwesterner."[24] The fate of the salmon is a clear example of the truth in his prediction.

Within the framework of the industrial economies and governmental institutions that dominate the watersheds of the Pacific Northwest, the search for a sustainable relationship between salmon and humans is a complicated mixture of politics, science, religion, and myths. For the past several decades, sustainability has been an elusive goal, and the wild salmon continue their slide towards extinction. All the power of science, all the organizational power of modern government, and all the wealth of the industrial economy have failed to find a way for humans and salmon to coexist in the watersheds of the Northwest. Historian Nancy Langston tells us that part of the failure stems from our failure to listen to and learn from the land and the rivers.[25] It is difficult to really see the land and rivers or hear what they are trying to tell us as long as what we hear and see is filtered through models of nature that cut us off from real learning and as long as the experts work within bureaucracies that impede such learning.

Side Channel 3
Visit to the River Machine

We would come to agree that henceforth no river should be appropriated in its entirety, nor be constrained to flow against its nature in some rigid, utilitarian straitjacket nor be abstracted ruthlessly from its dense ecological pattern to become a single abstract commodity having nothing but a cash value.

—Donald Worster[1]

Several years ago, I took I-84 out of Portland and headed up the Columbia River. I wanted to see firsthand what was being done to track the migration of juvenile salmon past John Day Dam. I drove under a thick layer of fog. It had crept inland, following the river all the way to the western edge of the gorge. The heavy fog cooled the lower half of the canyon. At the same time, I knew that, above the fog, the upper wall of the canyon was baking under a bright sun. Near the river, the branches of Douglas-fir, hemlock, and western redcedar combed water out of the moist air, creating a steady rain-like drip. I knew I would spend most of the day examining various kinds of machines and other forms of technology, so I pulled off the freeway and took a little walk in the woods and listened to the condensed fog drip onto the duff of the forest floor.

After I had passed the City of Hood River, green patches of irrigated crops dotted the dry brown meadows and hillsides, a sign that I had passed into the Cascade rain shadow. Further east, the rain shadow deepened so by the time I reached Rufus, Oregon, the Douglas-fir and cedar were gone. Sagebrush, a few scattered patches of the native bunch grasses, and the ubiquitous foreigner, cheat grass, covered the hills. At the little town of Rufus I pulled off the interstate and drove to a mostly empty strip mall and parked in front of the field office for National Marine Fisheries Service (NMFS). NMFS biologists were going to conduct the tour of the facilities and equipment used to monitor juvenile salmon migrating past John Day Dam.

After clearing security, we drove along the top of the dam to a place where several biologists wearing hard hats were working. While the tour was getting organized, I walked over to the upstream edge of the dam and looked over the guardrail. The dark water appeared to be dead still. Juvenile Chinook salmon swam along the face of the dam a foot or so below the surface. They swam back and forth like animals in a zoo pacing in their cage. Their large size said hatchery fish and their silver color said smolts on their way to the sea.

I knew that somewhere below these visible fish were many others following the suck of water into the turbines buried in the dam's concrete bowels. Many of them are screened out of the water and shunted into gate wells before they reach the rotating blades of the turbines. The gate wells are large shafts cut into the dam's concrete extending from the turbine screens to the surface of the dam. I looked down a gate well at the black water about thirty feet below and saw a solitary smolt swimming near the surface. I watched the little salmon for several minutes and thought about the journey that had brought it here and the dangers that still lay before it. The smolt was obviously stressed, swimming with its head partially out of the water. It appeared to be looking up at me while it swam in circles, lost and hopeless. I took out my notebook and wrote these lines:

> *At the bottom of a deep shaft*
> *Staring up from the gloom*
> *A solitary smolt*
> *Unaware it is doomed*

Juvenile salmon diverted into a gate well enter a trap that is periodically hoisted to the surface of the dam. Once the trap is free of the gate well, the juvenile salmon are flushed into a large plastic pipe that carries them into a building about forty feet away. Inside the building, NMFS biologists calm the salmon with an anesthetic and then they weigh and measure them. Many of the salmon both hatchery and wild carry a small piece of metal called a pit tag. The tag, which contains basic information about the young salmon, is retrieved electronically while the tag remains in the fish. Each tag is unique, so it identifies an individual fish and the hatchery it came from. Every time a juvenile salmon carrying a pit tag is captured at a dam its tag is read and the fish identified. Biologists use these data to track the salmon's progress down the Columbia and Snake rivers.

The whole operation—anesthetizing, weighing, measuring, and reading the tag and recording the numbers—is carried out with the mechanical efficiency of an assembly line. I am impressed by the intense use of technology to gather data on the juvenile salmon's downstream migration. Once a little salmon has given up all its information, it's released into another pipe that carries it back to the river on the downstream side of the dam. There, if the smolt is not eaten by a waiting gull or pike minnow, it continues its dangerous journey to the sea.

While our guide explained the process and technology used to gather the salmon's vital statistics, two more biologists drove onto the dam and parked nearby. They put on hard hats and set up electronic equipment that began emitting a series of beeps. The electronic device and its beeping aroused my interest, so I walked over to check it out. They were biologists from Oregon State University, and the beeping sound was coming from a radio receiver. The beeps were signals from small transmitters implanted in juvenile salmon. To me the beeps were just beeps, but to the biologists they were signals from several salmon somewhere in the slack water above the dam. The salmon were telling the scientists the story of their migration down the Columbia River.

The juvenile salmon moving downstream will spend most of the rest of their lives traveling, covering many hundreds, if not thousands, of miles. With their sophisticated equipment, the biologists at John Day and other dams study a few hundred feet or at most a few miles of that migration. They study a single link in the salmon's life history-habitat chain. They examine the same thin slice of space and time over and over and over again, trying to ensure the safety of the little salmon. They are a knowledgeable group dedicated to that goal. But near the end of the tour, I found myself wondering: Where is the context? Where is the ecological connection between salmon and river? Where is the connection to the rest of the salmon's life history? How is this one link in the life history-habitat chain connected to all the others? Without that context and without those connections, what I saw was a program to monitor the salmon's progress down a conduit to the sea, a conduit that was once a river—a conduit that was once a web of ecological relationships that embraced the salmon—and a river that once provided a context within which the salmon's life history played itself out.

That evening in a motel room I reflected on the day's events and what I had learned. I thought about all the truly dedicated scientists, working throughout the length of the Columbia River and into its tributary streams. I thought about the massive array of technology employed to achieve a seemingly simple goal—to get the small silver fish from the headwaters downstream to the river below the last dam.

But what really impressed me that day was a comment made by one of the biologists working at the John Day Dam. She said that she worried about the salmon's migration. When she had come to work at the John Day Dam several years before, the migration of juvenile salmon started in the spring and continued until September, after which the number of migrants tailed off. "But now," she said, "the migration is over by July or early August, more than a month earlier than just a few years ago." Among the biologists I talked to, she was the only one who seemed to be worried about the broader life history implications of the data being collected. When she stepped back from the narrow slice of the salmon's life that defined her job and took a broader look at the salmon's life cycle, she was able to see a real problem.

When Willis Rich studied the juvenile Chinook salmon in the Columbia River in 1916, he concluded that they migrated downstream most of the year, if not all year long. There was a diversity of migration times, which translates to life history diversity.[2] Life history diversity is one of the attributes that contributed to the salmon's phenomenal productivity before the rivers were developed. Salmon management based on the industrial model of the salmon's production system and based on a fragmented institutional structure functions most efficiently with less not more diversity. The problem of getting salmon past dams is reduced considerably if the period of migration is reduced from twelve to three months.

The visit to John Day Dam reminded me how hard it is to pay attention to the salmon, to the real animal in the midst of the massive use of technology, narrowly defined job descriptions, career concerns, and institutional politics. What I saw that day reminded me of the comment by Neil Evernden that "the object of science is theories not animals."[3] Too often the focus is on tools, technology, mathematical models, and theories. In the process we lose sight of the real animal. We have accepted as normal an "abstract nature—wildness objectified, and filtered through concepts, theories, institutions and technology."[4] The fragmented institutional structure and an industrial production system encourage this myopia. Among all the biologists I talked to that day at the dam only one tried to show how the piece of the salmon's life history she was studying fit into a broader context.

Chapter 4—Coda

Growing scientific evidence supports the notion that hatchery-caused problems cannot be ignored without further threatening the future of depleted salmon populations.
    —National Research Council[1]

The recently documented declines in Pacific Northwest salmon populations … indicate a breakdown in the west coast salmon management paradigm.
    —Eric Knudsen[2]

In the 1980s I attended a meeting to explore ways to incorporate the latest scientific understanding of salmon genetics and ecology into hatchery practices. The meeting started with a senior fish culturist explaining why the state's hatchery program couldn't change its practices and backing it up with pearls of wisdom such as the emphatic conclusion that salmon genetics, and especially the existence of genetically different, local stocks, was a bunch of bullshit. A coho salmon was a coho salmon and it didn't matter if it came from Alaska or California. I thought to myself, "This is going to be a long afternoon."

In the 1980s and later, fish culturists didn't publicly oppose scientific findings that challenged hatchery practices, but in meetings closed to the public they did not hesitate to state their opinions. In public, when confronted with evidence that their practices were not consistent with science, they hid behind a smoke screen of "we only do what the salmon managers tell us," which is partially true. Both fish culturists and managers adhere to the same salmon story. In closed meetings like the one I am about to describe, they used their considerable power within the agency to defend the status quo. But it was not the reluctance of fish culturists to accept new scientific information that lodged this meeting into my memory.

We were about halfway through the meeting when the word *product* fell out of the conversation. It fell onto the table with a loud clang and sat there glowing like a neon sign. The speaker went on—so many pounds of product at hatchery X, product at hatchery Y was diseased, and the product from hatchery A will be transferred to hatchery B, and so on. I had heard factory-produced salmon called products before, but on those occasions, the word just sailed by without making an impression. Today was different. The frequent use of the word *product* kept hitting me like repeated blows from a hammer. As I looked around the room, no one else seemed to notice that this garish word had intruded into our conversation.

On that day, the word *product* gave me new insight into salmon management. It reached down deep inside me and challenged my fundamental beliefs and assumptions about salmon management and what it meant to be a salmon manager. Products are what factories make and, if salmon are products, then are salmon biologists simply factory managers and peddlers of products?

For several days after that meeting the word *product* continued to fascinate me. I thought of assembly lines making products—skillful workers fastening fins, maxillaries, or eyes to the little salmon products. There were hatchery warehouses where boxes of product were stored, a little coho salmon in each box (batteries not included). I thought about how product could enhance our creativity. We could make Chinook salmon with fins like a '59 Cadillac. I even thought of a slogan to signify this achievement. The General Electric Company's slogan was "Progress is our most important product." Salmon managers could boast that in over a hundred years in practicing their profession, "Product has been our most important progress."

We were no longer discussing an animal with sixty million years of evolutionary history. Forget that the salmon is an animal with complex life histories. It's not important that their extended ecosystem ranges from the Gulf of Alaska to a mountain top in Idaho where an eagle feeds on a spawned out carcass. Do we really need to worry about the salmon's upstream migration and the mass movement of nutrients that feeds the whole ecosystem? On that day I realized that our salmon story had reduced this magnificent animal to a product and given it the same status as toasters, TV sets, and fishing poles.

What's in a word? There is a lot packed into the word *product*. Products are things we make. Maybe today, in a culture with an industrialized food chain, the possibility that we make salmon does not seem out of line. Where do salmon come from? Why, from fish factories, of course. Products are used and then thrown away when newer models become available. Products take a one-way trip from factory to junk yard with a brief stop in our homes. Products don't reproduce themselves. One generation of products doesn't pass on to the next generation the accumulated experience of evolutionary history. My coffeemaker (a product) doesn't stir up the same feelings and emotions in me as when I'm alone on a stream watching a pair of chum salmon spawning.

❧

A few years after that meeting I had the opportunity to ask Tom Jay, who is a writer, artist, and a wonderful teller of salmon stories, about the implications of making salmon a product. We talked about the origin and deeper meaning of the words *salmon* and *product*. The roots of the word *product* are the Latin *pro* and *ducere*—to lead forward. Deep inside of *product* is the linear, forward movement of the assembly line. But nature is not linear; it moves in circles and cycles—the cyclic appearance of seasons, the hydrologic cycle, and the life history cycle that brings each generation of salmon through the same points in the chain of habitats as the previous generation. Cycles and renewal are a fundamental part of the miracle of life and of the biophysical processes that support it. Look inside the word *product* and you will not see ecological cycles.

The word *salmon* comes from the Latin *salire,* to leap, which clearly captures the return of salmon to rivers and their upstream migration to spawn. Salmon the leaper creates in my mind a picture of a salmon jumping over falls, log jams, and beaver dams on its journey to close the cycle and start a new generation. Salmon may be a product to some, but when I look inside the word *product* I see no trace of the leaper.

Substituting the word *product* for *salmon* strips away the wonder and enchantment of the salmon's life history. *Product* severs the relationships between salmon and the landscape and converts nature to a warehouse of commodities. While we were busy transforming salmon into *product*, we failed to notice that we were at the same time diminishing their annual return to the rivers of the Pacific Northwest—killing our place-defining event. *Product* is another step in the erosion of our sense of place.

Once the salmon were incorporated into an industrial production system, it was an easy and logical step to reduce them to the status of a product. I'm sure some folks will argue that *product* is just an innocuous word without the deeper meaning that I am suggesting. But it is far from innocuous. It is a window into the heart of salmon management and at the same time the heart of the salmon's problem. A lot of ink has been used to describe the problems that hatcheries created for naturally reproducing populations of salmon. From my perspective all of the problems created by hatcheries—the genetic problems, the problems of mixed stock fisheries, and the loss of habitat traded for hatcheries—all of these problems are summed up in the word *product*.

Early in my career, I often referred to fish factories as neutral tools that only became negative or positive through use. I often used the example of a hammer, which is neither useful (to build a house) nor destructive (to smash a work of art) until some one takes it out of the toolbox and uses it. I was wrong. Fish factories are not neutral tools; they are part of a system of interacting and supportive or dependent technologies and economic activities. Fish factories need fleets of boats catching tons of pelagic fishes that are converted into bags of salmon feed. Dams and transmission lines supply the power used to pump water into hatchery ponds. Barges and trucks haul juvenile salmon past the dams, and trucks move juvenile salmon between hatcheries. The boats, barges, and trucks need oil wells and refineries for fuel. Chemicals used to treat the fish diseases that are common in hatcheries come from other factories. Hatcheries are part of a technological system with tentacles that reach into nearly every part of the economy. Once a technological system is established, those tentacles lay down strong roots and create an inertia that resists change. More importantly, technologies like fish factories are "culturally potent." According to Wolfgang Sachs: "Technologies shape feelings and fashion worldviews; the traces they leave in the mind are more difficult to erase than the traces they leave in the landscapes."[3]

Technology shapes our stories. In Chapter 2, "Salmon Stories," I showed how fish factories influenced how we think about salmon life histories and rivers. Think about how tools such as Google, Facebook, and Twitter have altered the nature of your relationships with other humans and the natural world. Contact with nature and other humans carried on largely through electronic devices enlarges our separation from nature, creates new filters on what we see, and even isolates us from real contact with other people. Will this lead to a greater estrangement from place? Will it increase our inability to see and experience nature and decrease our ability to comprehend the consequences of our dysfunctional behavior toward the place where we live and its natural attributes? Will our electronic relationship with nature make it easier or harder to change our salmon story? If it makes it harder to change the story, what lies ahead? Where will our current story take us?

One way to answer the last question is to project the assumptions and beliefs that underlie our salmon story into a vision of the future and then ask: Is this the future we want for ourselves and our children? I stumbled onto such a look into the future one day in the Fisheries Library at the University of Washington. While thumbing through a file of papers and reports, I came across one with the title "Plans and Details for a Comprehensive Development Program of Natural Salmon Rearing Areas in the State of Washington."[4] The document was prepared in 1964 by Milo Moore, who had served as director of the Washington Department of Fisheries. It describes a massive transformation of Washington State's bays and estuaries into engineered fish factories. As I sat in the library reading the document, I was alternately fascinated and revolted by the extreme techno-utopian vision unfolding in the plan's seventy-three pages.

The plan starts with a description of the Pacific salmon crisis and correctly links the collapse of abundant salmon runs to excessive harvest and habitat alteration. It also endorses the earlier belief about the role of predation in decline of salmon. According to the plan, predators are a big problem that management agencies are not adequately addressing.

Then the real purpose of the plan is revealed. Millions of eggs are wasted by allowing salmon to spawn in rivers with degraded habitat. These wasted eggs could be put to productive use, if enough money were provided to artificially spawn the salmon. This last point sounds a lot like Livingston Stone's statement nearly one hundred years earlier:

Nature, perhaps more aptly speaking, Providence, in the case of fish, …
produces great quantities of seed that nature does not utilize or need. It
looks like a vast store that has been provided for nature, to hold in reserve
against the time when the increased population of the earth should need it
and the sagacity of man utilize it. At all events nature has never utilized this
reserve, and man finds it already here to meet his needs.[5]

The author of the paper I was reading complains that millions of dollars spent on fish research had failed to halt the salmon's decline, and that this failure indicates the need for a "more positive effort" to increase the salmon's abundance. This characterization of research ignores two important historical details: 1) For several decades, this "more positive effort" was exchanging salmon habitat for fish factories faster than research could gather new information on the salmon's ecology; and 2) salmon management had been slow to incorporate new information, especially new information that contradicted or challenged the basic assumptions in the salmon story. These introductory comments indicated that this was not an innovative or new approach, but a look at the future through the lens of the old story.

For $17.3 million, the program would have implemented twenty-two projects producing 189.3 million migrant-sized juvenile salmon and 124,000 oyster strings. This "more positive effort" would have dammed, engineered, and converted many of Washington State's bays and sloughs into large salmon factories. The plan for the Dungeness Bay fish factory was typical of many of the other proposed projects. A "leaky dam" across the mouth of Dungeness Bay would impound twelve hundred acres, converting the bay into a rearing pond for twenty-five million juvenile salmon. I'm not sure what the term *leaky dam* means. I assume it meant that the dam would allow some movement of water, but not the fish, in and out of the bay during tidal exchanges. The concurrent culture of oysters and Dungeness crabs would add benefits to the project. The plan predicted that these "improvements" would stimulate the construction of summer homes around the bay, including the Dungeness Spit, boosting the project's recreation value. Lost in this promise of enhanced economic value was the fact that the spit and a part of the bay were designated a national wildlife refuge in 1915. Would a massive fish factory and homes on the spit conflict with the refuge? The author does not answer that question. Under the enhancement plan, at least some of the birds that frequent the bay today and are protected by the refuge would have been eliminated by "[a] full scale fish and bird predator control program."

The plan called for the conversion of Sequim Bay, also on the Olympic Peninsula, into a giant hatchery pond. A sill across the bay's narrow mouth made a leaky dam unnecessary. The sill combined with several artificial feeding stations would have kept juvenile salmon in the bay until they had the natural urge to migrate. Fish traps would continuously remove salmon predators. The plan included an enhanced recreational fishery on marine species. To create this fishery, two boats equipped with otter trawls would catch marine fish in other areas of Puget Sound or the Strait of Juan de Fuca and stock them into Sequim Bay. Some of the fish stocked in the bay would probably be the same predators on juvenile salmon as were removed by the fish traps.

The massive conversion of natural bays into engineered salmon factories was proposed in the mid-1960s, at a time when hatcheries were benefiting from a wave of renewed enthusiasm. Managers were witnessing dramatic increases in salmon abundance, creating rapidly expanding sport and commercial fisheries. The increase in abundance of factory-produced salmon was largely attributed to better disease control and more nutritious feeds. These improvements produced healthier smolts, which allowed fish culturists to explore new hatchery practices, such as longer rearing of the juveniles before release. All of this led to higher survival of salmon, both in the hatchery and after release. Hatcheries were finally delivering on their promise to enhance or at least maintain the supply of salmon. Trading habitat for salmon was a good deal after all. It was during this renewed belief in the effectiveness of the industrial production of salmon that the plan for a massive development of Washington State's bays was unveiled.

The salmon story eliminated consideration of other factors that might have been contributing to the salmon bonanza. And there were other factors. Better hatchery practices did contribute to the increased abundance, but most of the increase was the result of changes in the ocean environment. The myopia imposed by the salmon story rendered the ecological changes taking place in the ocean invisible. In 1976, the salmon bonanza abruptly came to an end.[6] The ecological factors that had earlier improved the ocean survival of salmon changed and salmon abundance plummeted. Fish factories were unable to reverse the decline.

Unfortunately that's not the end of this story. In order to harvest the large number of factory-produced salmon, managers overharvested the wild salmon in mixed stock fisheries. This overharvest set the stage for the dramatic collapse of wild salmon in the 1980s and '90s, leading to the listings under the federal ESA.

⌒·⌒

The grandiose plan was not implemented. Maybe it was too expensive. Perhaps the change in ocean conditions in 1976, and the decline in salmon abundance that followed, undercut the enthusiasm for fish factories. The events following 1976 probably cast too much doubt on the plan's ability to deliver the promised benefits. By the mid-1970s fishery scientists began to question the wisdom of relying on artificial propagation to maintain the supply of salmon. The continued decline in salmon abundance sparked debates about the use of fish factories and the impact of artificially propagated fish on wild salmon populations. The debate continues today, although it has made little difference in salmon management's conceptual foundation.[7] Although the massive conversion of Washington State's bays and estuaries into salmon factories was not implemented, the vision of a simplified production system based on industrial rather than ecological principles is still alive and it is still trying to push salmon out of their rivers and incorporate them—much like chickens, hogs, and cattle—into the industrial food chain. Today, the industrialization of salmon production is proceeding in smaller, less grandiose steps. The steps may be smaller but they are on the same trajectory as the plan to massively change the bays and estuaries of Washington. What follows are descriptions of some of those steps.

Sea Ranching

The increase in ocean survival of salmon in the 1960s caught the attention of private business interests. They envisioned a salmon ranching enterprise modeled on the open-range cattle operations of the late nineteenth century. Salmon ranching would be easier, because adult salmon return to their home stream, eliminating the need for an expensive annual roundup. Corporate proponents of this scheme saw rivers of silver fish returning to their salmon factories, where they would convert the silver to gold. The Oregon legislature saw the corporate interest in fish factories as a way to boost the industrial output of salmon. It authorized private salmon ranching operations for chum salmon in 1971 and, in 1973, amended the original statute to include Chinook and coho salmon. The Weyerhaeuser Corporation quickly invested $40 million in two sea ranching operations, one in Coos Bay and the other in Yaquina Bay.[8] Other corporations also took advantage of the new law.

The private sea ranches relied on the same simplified model of the salmon's life history that underlies the public management of salmon. Private fish factories located on or near the coast spawned adult salmon, incubated the eggs, and reared the juveniles within the private fish factories. The juvenile salmon were then released into a stream that carried them directly to a bay, estuary, or ocean. In some cases this stream was an engineered channel flowing out of the factory. After grazing the ocean pastures, adult salmon returned to the stream where they were released and entered a processing plant, where they were converted into a product. The whole process simplified the salmon's life history and industrialized their production to a greater degree than any of the public hatchery programs. Private sea ranching, its proponents declared, would lift the salmon out of the primitive hunter-gatherer economy and place them in the modern economy and its industrial food chain. But this claim of a modern, new approach to salmon production was false. This scheme was a slightly modified version of the salmon story adopted in the nineteenth century. The corporate owners of sea ranches did not see that their scheme already had a century-long history of failure.

The proponents of private sea ranches argued that habitat degradation reduced the capacity of rivers to produce wild salmon, which was true. They reasoned that fewer smolts migrating to sea left vacant habitat in the ocean, which they would fill with salmon from their private fish factories. This assumption meant that the proponents of private fish factories believed the ocean was not an ecosystem with a complex web of ecological relationships, but a collection of fixed niches that each species occupied independent of the rest of the ecosystem. According to this vision, when a species declined in abundance it created vacant capacity in its niche, which would remain vacant until filled by the same species—the vacancy would not be invaded by other species. Proponents of sea ranching made the assumption that competition and other ecological interactions between species did not exist in the ocean. Once again a simplified model of an ecosystem was created by ignoring ecological complexity. This extension of the old salmon story to the ocean, even though scientifically flawed to the point of being ridiculous, was so powerful it lured private industry to invest millions. If there were any doubt that the old salmon story was still influencing policy, sea ranching was proof measured in dollars.

The simplification of the salmon's production system, which began in freshwater, was now extended into the ocean.

Weyerhaeuser, along with other corporations, was unwittingly conducting an expensive experiment, an experiment that would test the validity of a key assumption embedded in the salmon story: natural, salmon-sustaining ecosystems can be replaced by simple industrial processes. By 1992 the results of the experiment were in. In that year, the last private sea ranch on the Oregon coast closed. Private corporations had bet several million dollars on the belief that salmon factories were economically viable, and they lost. Their failure should have exposed a major weakness in salmon management's conceptual foundation, but it didn't.

Feedlot Aquaculture

Just as the disastrous winters of 1881 and 1882 helped bring the open-range ranching of cattle to an end, the change in ocean conditions in 1976 reduced salmon survival and wrung the profits out of sea ranching. Private sea ranching failed, but private salmon farming did not disappear. To continue the comparison with cattle production, salmon farming moved from the open range to feedlots or high-density rearing of salmon in net pens anchored in marine waters. Feedlot aquaculture put salmon under human control throughout their entire life cycle—a further and perhaps the ultimate simplification of salmon-sustaining ecosystems. In Chapter 1, I said that the functioning of the market is replacing the return of the wild salmon as the place-defining event in the Pacific Northwest. Feedlot aquaculture survives only because it feeds that market with a product. Every time I hear that "we need market solutions to environmental problems," I cringe and wonder. Do the folks making those statements understand the market's role in the collapse of wild salmon?

Feedlot production of salmon increased rapidly and is a global industry today, with operations in Norway, United Kingdom, Canada, Chile, and the United States. By 1990, pen-reared fish dominated the world trade in salmon. British Columbia and Washington State had 121 and nine salmon farms respectively in 2003.[9] Net pens give the illusion that they isolate the farmed fish from the surrounding waters in the bay or fjord where they are located. However, they are not isolated and they are not benign. Salmon feedlots are a serious threat to the wild salmon populations and to the commercial and sport fishermen and the local communities that depend on the wild salmon. Many of the threats posed by feedlot aquaculture are described in an excellent book, *A Stain Upon the Sea*.[10]

Feedlot aquaculture tears into an aquatic ecosystem like two mismatched gears being forced together. If only the placement of feedlots in a pristine bay signaled the serious problems they create like the loud grinding and screeching of mismatched gears, the problem would be obvious and the debate short. But there is often a pristine silence in the waters degraded by feedlots. The dying and dead wild salmon are invisible to all but the most diligent observers. So the debate rages while wild salmon populations are degraded to the point of total collapse or extinction. The debate between the proponents of industrial food chains and the advocates of wild salmon will never be resolved, because the folks concerned about wild salmon and those pushing feedlot aquaculture have such widely different stories or worldviews that at best effective communication is very difficult.[11]

In 2007, I traveled to the center of that debate and met a remarkable woman and a small group of people trying to save their wild salmon and their community. The meeting took place in the Broughton Archipelago, British Columbia. I hitched a ride to Echo Bay Marine Laboratory on Howard Pattison's boat out of Telegraph Cove. I was there to attend a gathering of salmon people who were discussing the threat to wild salmon posed by the nearby feedlots. Howard's boat crossed Johnstone Strait and moved through a beautiful land and seascape that had the look and feel of pristine salmon country. The Broughton is the unlikely place where the clash between wild salmon and their dependent human communities and the industrial food chain with its feedlot aquaculture has come to a head. At Echo Bay I met Alexandra Morton, the woman whose untiring work has shined a light on the dark side of salmon feedlots. She has exposed the fact that they kill wild salmon, and ultimately, if they continue in their present mode of operation, will kill the Broughton's fishing community.

Alex saw her first salmon feedlot in 1987, and as she watched the seemingly harmless pen being towed past her window, it gave no hint of the disastrous effects it would soon have on wild salmon. Later, as she reflected on that day, she said, "I wish now we had done everything possible to stop that first farm. But we had no idea."[12] After the arrival of the salmon feedlots, Alex, who was studying whales in the Broughton, began noticing and documenting things that disturbed her. The feedlots were being located in areas used by wild salmon, the very places that the government said would be off limits. Strange disease outbreaks in wild salmon populations appeared to be related to disease problems in the feedlots. Whales were being driven from their historical habitats by acoustic devices employed to keep predatory seals away from the net pens. Non-native Atlantic salmon were escaping from the feedlots and were being caught by fishermen. Would this alien species become competitors with wild salmon and steelhead for spawning and rearing habitat? Tons of waste generated by the feedlots was polluting adjacent waters. And, possibly the biggest problem, unresponsive bureaucrats whose adherence to the old salmon story and its fish factories was so in line with the salmon feedlots that they could not see the problems, or worse, didn't seem to care.[13]

Alex also identified the life-threatening problem of sea lice, small, naturally occurring parasites that attach themselves to salmon and feed on their skin, muscle, and blood. They are common on adult salmon and a few small parasites on the large body of an adult salmon are considered benign. However, a heavy infestation of sea lice can kill a small juvenile salmon. Normally, the juvenile and adult salmon travel the marine pathways to and from their home rivers at different times. This reduces the chance that sea lice can be transferred from the adults to juveniles. Salmon feedlots destroy this natural separation. The large number of infected salmon concentrated in feedlots amplifies the density of sea lice in the surrounding water. Juvenile salmon pick up sea lice as they migrate past the infected feedlots. Mortality rates of the juvenile pink salmon infected with sea lice are commonly 80 percent. In the Broughton, where Alex does her work, sea lice from feedlots may be the proverbial straw that breaks the wild salmon's back. In a paper published by Alex and her colleagues in the prestigious journal *Science*, the authors concluded that continued exposure to swarms of sea lice from feedlots will cause local extinctions of wild pink salmon and a 99 percent reduction in abundance in four generations.[14]

The industrial production of salmon, either in feedlots or conventional fish factories, contributes to the decline of wild salmon, which, in turn, stimulates the call for more factory-produced fish to boost declining numbers. This is called progress by some biologists. They try to spin the disaster stemming from their policies as the inevitable march of progress. Wild salmon, they say, are quaint antiques from an earlier, primitive era of hunter-gatherer economies, or they give wild salmon a negative spin by calling them feral fish. This selling of the status quo reached the surreal one day when I was told in all seriousness that feral salmon in a river were a disease threat to fish in a nearby fish factory and so the feral fish should be destroyed. The power of the salmon story to rationalize such thinking is truly amazing. Is it progress? Is it inevitable? Is it new and modern? Feedlots and fish factories are modern production systems in appearance, but their heritage is same old story.

A new, disturbing chapter is being added to the salmon story: fish factories will, in addition to serving markets, save wild salmon from extinction and contribute to their recovery. In other words, fish factories are the solution to the problem they helped create.

Supplementation and Conservation Factories

The modern environmental movement gained momentum in the1960s with the publication of Rachel Carson's *Silent Spring*.[15] Concern for the state of the environment continued to grow over the ensuing decade, leading to landmark legislation that created the Environmental Protection Agency (1970) and other legislation such as the National Environmental Policy Act (1969) and the Endangered Species Act (ESA) (1973) and led to the first Earth Day in 1970. Against this background, the salmon's long-standing decline mobilized public and political support for large recovery programs such as the Salmonid Enhancement Program in British Columbia, the Lower Snake River Compensation Program, and the Northwest Power and Conservation Council's Fish and Wildlife Restoration Program for the Columbia River. I discussed each of these in my previous book, *Salmon Without Rivers*. Those recovery programs were implemented with the best of intentions. However, good intentions are not, in themselves, sufficient to achieve salmon recovery. The underlying conceptual foundation and its assumptions about how nature works determine whether the program will achieve its goal or contribute to the problem.[16] Consistent with the assumptions about nature in our salmon story, a common theme in all those restoration programs is the fish factory.

By the 1970s, it was becoming hard to ignore the century-long failure of hatcheries to maintain the supply of salmon. To continue to sell recovery programs based on an industrial production system, managers put a new face on it; they simply attached the word supplementation or conservation to the word hatchery. In many rivers, salmon habitat has undergone a century or more of degradation, in part because it was traded for fish factories. Wild salmon in many of those streams are protected by the federal ESA. This is where conservation hatcheries come in. They will help rebuild wild stocks by boosting the number of salmon while "lessening the genetic and ecological impacts of hatchery releases on wild fish."[17] In recent years, fishery scientists have published operational guidelines that, if followed, would make fish factories live up to their new-found conservation role, but to my knowledge no hatchery program has fully implemented them.[18]

Supplementation and conservation hatcheries have similar goals and constraints regarding genetic and ecological impacts.[19] Their underlying rationale is based on the higher survival of juvenile salmon in the protected habitat of the fish factory, discussed earlier in Chapter 2. This early survival advantage means that a pair of salmon spawning naturally in a river will produce fewer juvenile salmon migrating to sea than a pair of salmon spawned in a hatchery. From a strictly numerical perspective, taking salmon from the natural environment and spawning them in the fish factory produces more smolts migrating to sea. (On the other hand, when released from the hatchery, artificially propagated juveniles have lower survival to adult than their wild cousins.) Then if the returning adults are allowed to spawn naturally it should, according to supplementation's rationale, result in an increase in natural production. Some biologists refer to this as "jump starting" a depleted salmon population. Equating a depleted salmon population to an engine with a dead battery not only exposes supplementation's industrial roots, but it exposes thinking that is largely devoid of ecological reality.

Successful supplementation depends on completion of a critical parallel activity. The problem (overharvest, dams, or habitat degradation) causing the decline in abundance must be corrected prior to or at the same time the depleted population is supplemented with factory-produced fish. Some supplementation programs have attempted to comply with this assumption, but supplementation is being implemented in the Columbia River on a large scale without a corresponding level of attention to the factors limiting salmon abundance. The Independent Science Advisory Board recognized the contradiction when it said, "Supplementation is a tool that has been proposed for maintaining salmon populations in the areas where they would be marginal or nonexistent under otherwise existing habitat conditions. Supplementation per se does not improve natural habitat conditions. Under adverse habitat conditions supplementation cannot be expected to produce offspring that can sustain themselves without assistance."[20] Supplementation programs that continue to release artificially propagated juveniles without also fixing the cause of depletion are really nothing more than conventional hatchery programs that continue the practice of trading habitat for hatcheries.

Once a stream is supplemented with factory-produced juveniles, degraded habitat, mainstem dams, and overharvest will reduce the number of salmon returning to the home stream to spawn. Those sources of mortality can reduce or eliminate any benefits from supplementation. To circumvent mortality between their release from the hatchery and their return as adults, some supplementation programs never release the salmon from the hatchery; those salmon spend their whole life in a fish factory. This approach to supplementation is called captive brood, which theoretically ensures a larger supply of mature adults available for spawning. The progeny of the adults that spent their entire life in a fish factory are then used to "jump start" depleted salmon populations. Captive brood programs cut the salmon off from their natural life history and their ecological and evolutionary relationships. The salmon in captive brood programs are totally cut off from any, to use Gary Nabhan's words, ecological companionship. Once again, this is being done in the name of progress, before the consequences of such a drastic action are fully understood. Like the salmon feedlots, captive brood is a logical extension of the status quo.

The failure to correct the problems causing salmon decline, including the mainstem dams, before investing in supplementation programs produces embarrassing outcomes. For example, the Independent Economic Advisory Board for the Northwest Power and Conservation Council examined the operation of seven fish factories in the Columbia Basin. The hatcheries were located throughout the basin from the lower river below the dams to the upper basin above several dams. The average cost per adult fish harvested ranged from $14 per fish for a coho salmon planted in Youngs Bay near Astoria and below all the dams, to $23 per fish for fall Chinook released from Priest Rapids Hatchery and a whopping $68,031 per fish for Entiat Hatchery spring Chinook.[21] Entiat Hatchery is above Priest Rapids Dam in the upper Columbia River.

Select Area Fisheries Enhancement (SAFE)

The eighty-year-old war over the allocation of salmon harvest in the Columbia River bubbled to the surface again recently and was reported in articles and an editorial in the Portland *Oregonian*.[22] Unlike most of the salmon allocation wars of the past, which were between the users of different kinds of commercial fishing gear—trap fishermen vs. gillnetters and gillnetters vs. trollers—the current clash is between the sport and commercial fishermen. Sport and commercial fishers have been fighting to increase their catch allocation of salmon for a long time, but this time a new wrinkle was added: the plight of the sport fishing industry, the industry that sells sport fishermen all the gadgets they need to catch a salmon. The financial problems of one of the region's leading retailers of fishing equipment and one of the largest manufacturers of sport fishing boats added heat to the debate. It reminds me of something a wise old biologist told me early in my career. "When the last salmon swims up the Columbia River, any crying over the impending extinction of the species will be drowned out by the battle over who gets to harvest it."

The controversy boils down to this: to many Northwesterners, salmon are dollars, and the current debate is part of a long-standing struggle to control the invisible hand of the market so that more of the salmon dollars go into the pocket of this or that special interest. Managers have been drawn into this conflict so deeply that, as Tim Bowling, a writer who grew up in a family that fished commercially in the Fraser River, has observed, the salmon are considered a chunk of cash to be divided up by bureaucrats. I have a great deal of sympathy for the retailers, the guides, the commercial fishermen, and all of their families. I have sympathy for the sport fishermen, among whom I count myself. I consider the loss of a salmon fishery, whether commercial or sport, as tragic as the loss of a salmon population. Our culture and the salmon are too intertwined to see it any other way. In its reporting on the controversy, *The Oregonian* focused on the economics of the current battle, but it is a story that begs for a deeper look.

In spite of the billion-plus dollars spent on salmon recovery in the Columbia River in the last twenty-five years, the ecosystem that once nurtured annual salmon runs of ten to fifteen million fish is impoverished and its supposed replacement—an industrial production system—is mired in failure. Even though the Northwest Power and Conservation Council's goal is to recover only about 30 to 50 percent of the historical salmon abundance, the achievement of that modest goal would produce an annual run of about five million adult fish. That number of salmon would surely cool the fires of the harvest war. However, in the first twenty years of the Council's salmon recovery program, salmon runs fluctuated between a low of seven hundred fifty thousand fish to a high of three million fish. To its credit *The Oregonian* recognized that salmon runs have not met expectations, but in a way that makes it seem as though the failure of recovery efforts and the harvest war are separate problems instead of two sides of the same coin. *The Oregonian* should have acknowledged that the economic health of the salmon fisheries and the industries and businesses related to those fisheries is first and foremost dependent on the ecological health of wild salmon and their rivers.

One proposed solution to the current battle in the salmon harvest war is worth a few comments. The official title of the program is the acronym SAFE, for Select Area Fisheries Enhancement. I won't go into its scientific shortcomings. Anyone interested should read the scientific review of the program by the joint Independent Scientific Review Panel (ISRP) and the Independent Economic Analysis Board (IEAB).[23] The science panels raised red flags of caution, which the current SAFE proposal glosses over. Basically the SAFE program would increase the production of fish factories. The increased output would be released in areas of the Columbia River that are off the main channel. When the factory fish returned as adults to the point of release they would be harvested in the commercial gillnet fishery restricted to those release places. Theoretically, this would eliminate the commercial harvest on all the other hatchery and wild salmon following their normal migration route up the river and give the sport and Native American fishermen a larger harvest quota.

SAFE illustrates one attribute of the status quo: Don't fix the cause of the conflict between sport and commercial fishermen (the failure of recovery programs), attack the symptom (too many fishermen chasing too few fish). And don't just attack the symptom, but attack it with the same technology that contributed to the problem in the first place. If implemented, SAFE would be another example of a management action responding solely to the criteria of economism at the expense of stewardship. SAFE treats salmon as simple widgets produced in a factory and ignores the salmon's relationship to place. The ecologically placeless factory fish are planted according to economic criteria while ignoring the ecological and evolutionary implications. The plan is touted as a win-win solution. Sport fishermen and commercial fishermen win something. However, SAFE's proponents forget there are three parties in this controversy: the commercial fishing industry, the sport fishing industry, and the salmon. The plan is really a short-term win-win for the sport and commercial fisheries, but because the salmon are losers in the deal, all three parties will lose in the long term.

The irony of this is that the same approach that helped impoverish the commercial fishery is now proposed as the way to let that fishery capture a small shard of its past glory. SAFE is a new wrinkle on the old salmon story, and it amounts to rearranging the deck chairs on the good ship status quo as it heads straight toward the rocks of extinction. Using words like enhancement and salmon recovery and the acronym SAFE, the program builds a positive frame to hide the fact that it is really the same failed story.

If you have read this far in the book, you should realize that the SAFE program is detached from an ecological foundation. It is another instance of a management approach that views every problem as a nail to be hit with the hatchery hammer. Salmon management has a dual obligation: to supply salmon for the benefit of those economically tied to sport or commercial fisheries and to ensure the ecological health of the salmon through effective stewardship. SAFE focuses on the former and ignores the latter. When programs like SAFE are proposed, politicians and the public must insist that they contain a stewardship element to balance the program's economic purposes. For example, along with the implementation of SAFE, salmon managers could decide it was an opportune time to set escapement levels for all the extant wild populations in the Columbia River. Then set up a monitoring program to ensure that those target escapements were met. This would add a stewardship component to a plan that definitely needs one.

Of course, escapement targets and monitoring programs should have been set and implemented a long time ago. In fact, my mentioning them here as part of a proposed solution to a salmon allocation war is not a new idea. In an earlier allocation war in the Columbia River, fish wheels were eliminated in Oregon (1927) and fish wheels, traps, set nets, and seines were eliminated in Washington (1935). In 1948, biologists with the Oregon Fish Commission analyzed data on salmon escapements for nineteen years (1928-46) to determine if they improved after the elimination of several kinds of gear. Here is part of what they concluded:

> *It appears that the elimination of any one type of gear on the Columbia River has served only to increase the catch of other gears rather than increase the escapement. ...This is interesting in view of the fact the elimination of fixed gear (fish wheels, traps, seines, and set nets) in 1935, from the Washington side of the river was undertaken presumably to reduce the catch and increase the spawning escapement.*[24]

Over sixty years ago we learned that programs with a noble conservation purpose can go astray unless the public is vigilant in insisting on a balance between economics and stewardship.

Salmon are part of the commons; they belong to all of us, and we all have an obligation to ensure their persistence. The fact that we have given stewardship responsibilities to states and federal agencies does not diminish our obligations to conserve the commons. National parks are also part of the commons. What if the National Park Service ignored the dual responsibility implied in its mission and decided to focus almost entirely on maximizing the parks' economic return. The result would be a logged-off Disneyland, where people would participate in a virtual nature experience using electronic gadgets and 3-D theaters.

Salmon ranching, salmon farms (feedlots), conservation hatcheries, supplementation, "jump starting," captive brood, and SAFE are concepts with roots in the simplified industrial production system. At first they appear to be new and innovative, but upon close examination, it's clear they are a derivative of a story that has racked up more than a century of failure. However, an objective examination of those myths, beliefs, and assumptions is difficult to carry out. Equating the myths and assumptions to a lens—a cultural lens—helps explain why it is so difficult. Imagine that you wear glasses and you cannot see clearly without them, but the glasses are broken—a screw has fallen out or maybe the frame is cracked. You take your glasses off to examine them and quickly discover that you cannot both look through the glasses and examine them at the same time. The same problem arises when we try to examine the lens through which we see the world around us, the lens created by our cultural filters. It is difficult to look though the lens and examine the validity of the vision it gives us at the same time.[25] It is so difficult that it is rarely undertaken, and so the proponents of supplementation in its various forms cannot see that they have pinned their hopes for salmon restoration on the same assumptions, beliefs, and myths that created the salmon's problem.

Here is a brief history of the terminology used to periodically reinvent fish factories. As you read this, keep in mind that prior to the 1930s or 1940s the massive harvests of salmon were composed entirely of wild fish. In the late nineteenth century, salmon managers believed they could *enhance* the supply of salmon beyond natural limits with an industrial production system. They wanted to replace ecological processes that they did not understand with industrial processes that they could control. They did not understand the consequences of that control. Hatcheries failed to achieve the goal of increasing salmon abundance beyond natural limits and salmon began to decline because of excessive harvest and habitat degradation.

By the 1920s the goal of hatcheries shifted. Now fish factories were going to reverse the decline and *maintain* the supply of salmon. Salmon continued to decline. In the 1940s, mainstem dams and river development emerged as a new threat to the salmon. Without clear evidence that fish factories had achieved their previous goals, managers once again turned to them. Now the goal was to *mitigate* the negative effects of dams and development. Through the 1940s, '50s and '60s mitigation was synonymous with fish factories. Salmon continued to decline.

The depleted state of the salmon reached critical levels in the late 1970s and led a decade later to petitions to list wild salmon under the federal ESA. Salmon managers now faced a new problem: how to prevent extinction and increase salmon abundance to levels that would remove them from threatened or endangered status. Once again fish managers tacked a new name and a new mission on the same old fish factories and sold them as the solution to the new problem. This time fish factories were dressed up in terms like *supplementation*, *conservation*, or *captive brood*. And so here we are in the twenty-first century chained to a tool that has a record of failure, but has an uncanny ability to adapt and reinvent itself for every situation. This scenario reminds me of the old saying: if the only tool you have is a hammer, then every problem looks like a nail.

By the end of the twentieth century, salmon research increased our understanding of salmon-sustaining ecosystems and exposed a growing gap between the new scientific understanding and management practices emanating from an outdated conceptual foundation. Eventually, the gap was so obvious that it triggered calls for hatchery reform.[26] A few of these reform efforts have actually attempted to re-conceptualize the role of hatcheries from an ecosystem perspective.[27] But the old story has resisted and deflected real examination and accountability for over a century. Its proponents are good at avoiding accountability. So the old story remains hidden, buried deep in the culture of the salmon management institutions, supported by politicians looking for easy answers and those who want to continue degrading salmon habitat.

I am not optimistic that the current reform efforts will actually change the status quo. To the leaders of salmon management institutions, the prospect of loosening their ties to the old salmon story is fraught with political danger. Such a move would force them to confront the ecological complexity and uncertainty that the old story has allowed them to largely ignore. Confronting ecological reality would also force them to confront the bureaucratic boundaries that fragment the management of ecosystems. The stubborn adherence to an industrial production system that is destroying salmon-sustaining ecosystems strongly suggests that salmon management institutions and their leaders have lost sight of their mission.

This raises two questions, the answers to which will take up the remainder of this chapter: What is the mission of salmon management agencies and how could the leaders of those institutions lose sight of their agency's purpose? In his book, *The Institutional Imperative*, Robert Kharasch makes a distinction between an institution's ostensible purpose and its real purpose. The ostensible purpose is the laudable, publicly declared mission of the institution. The real purpose is what the internal machinery of the institution is set up to do. In many cases, the internal machinery of an agency is actually detrimental to its ostensible mission.[28] According to its website, the mission or ostensible purpose of the Oregon Department of Fish and Wildlife (ODFW) is:

To protect and enhance Oregon's fish and wildlife and their habitats for use and enjoyment by present and future generations.

The Fish Division expands on this mission and restates it as:

The mission of the Oregon Department of Fish and Wildlife (ODFW) is to protect and enhance Oregon's fish and wildlife and their habitats for use and enjoyment by present and future generations. The Department is charged by statute (ORS 506.036) to protect and propagate fish in the state. This includes direct responsibility for regulating harvest of fish, protection of fish, enhancement of fish populations through habitat improvement, and the rearing and release of fish into public waters. ODFW maintains hatcheries throughout the state to provide fish for program needs.

Salmon management agencies in other Northwest states have similar sounding missions, especially with regard to resource protection. (See their mission statements at the end of this chapter.) If the status of salmon today is compared to the ostensible purpose of the agencies it leads to the inescapable conclusion that salmon management is failing to accomplish its mission. In the Pacific Northwest, salmon are extinct in 40 percent of their historical range and the status of salmon in much of the remaining range is so precarious that they are protected under the federal ESA.[29] How could anyone argue that the salmon have been protected or enhanced? How could anyone argue that the trajectory of salmon abundance over the past century shows a concern for future generations?

For Oregon, the Fish Division's expanded mission statement exposes its internal machinery. The juxtaposition of "to protect and propagate" comes close to an oxymoron, in that salmon factories have been an impediment to the salmon's protection. The internal machinery of the fish and wildlife institutions is designed to implement the salmon story and its industrial production system. The real mission, as opposed to the ostensible purpose, is to run fish factories and maximize the economic return of the commodities those factories produce. Anyone who doubts my characterization of the real purpose of the ODFW's Fish Division should read the budget bid sheets submitted in 2012 to justify their request for state funding. The Fish Division uses only economic metrics to demonstrate the performance of its programs. Those economic data are used as a proxy for the heath of the resources the Division manages. You couldn't ask for clearer evidence of what the internal machinery of the Fish Division is geared to achieve.

For over a century, salmon abundance in the Pacific Northwest has, with some variation, followed a downward trajectory. Some populations are already extinct, while others are precariously close to the edge of an extinction spiral. Over the same period, especially during the past seven or eight decades, salmon have been "enhanced" through a series of management and recovery programs. How is it possible that the promised outcome of enhancement and the reality of depletion and extinction have been able to coexist over so many years? Why didn't the disparity between the promise and the reality lead to calls for accountability? How do the staff at an agency that was created to achieve a laudable purpose develop patterns of behavior that fail to achieve the stated mission? How do individuals within those agencies justify actions that are not supportive of, and may even conflict with, the institution's ostensible purpose?

Those questions bothered me for a long time. I couldn't explain the contradiction until I talked with David Bella. Dave is a professor emeritus of civil engineering at Oregon State University and has had an interest in and has published valuable research on the functioning and malfunctioning of institutions. At the heart of Bella's work is something he calls the "behavioral context." Don't let this academic-sounding term put you off. It offers real insight into this hidden part of the salmon's problem.

According to Bella, contrary and harmful outcomes can emerge from individual behaviors that are seen within an institution "as competent, normal and even commendable."[30] Within salmon management institutions, competent and dedicated individuals are working hard, but the outcomes of their work—extinction and depletion of wild salmon—are contrary to expectations and contrary to their institution's mission. This problem is not limited to fish and wildlife management agencies.

In a postscript to the recent near meltdown of the U. S. economy, which is relevant to the theme of this book, Alan Greenspan, in testimony before the U. S. Congress regarding the economic crisis, admitted the conceptual framework that had guided his treatment of the economy for forty years had a flaw.[31] If an administrator of a salmon management institution made a similar forthright admission, it would be an important step toward building sustainable relationships between salmon, people, and place.

Bella's behavioral context explains how the work of individuals deemed competent and normal produces outcomes contrary to an agency's purpose—the performance norms shared by all the members of the institution are simply taken for granted. Bella's behavioral context is similar to what Robert Kharasch called the internal machinery of an institution. The first three chapters of this book described the behavioral context of salmon management institutions. The context consists of cultural filters (Chapter 1) that determine what we see in the world around us and how we behave towards it. Those filters helped create our salmon story (Chapter 2) and the industrial production system. The story is reinforced by the fragmented management of ecosystems (Chapter 3). This behavioral context produced a century of results contrary to the mission of fish and wildlife agencies. Behavior that persists in the face of a century of failure is a verification of Bella's belief that there are few things that can hide and sustain a problem as well as the appearance of normalcy,[32] and working within the constraints of the salmon story has the appearance of normal behavior.

Bella uses a flow chart to show how the behavioral context produces outcomes contrary to expectations. I constructed one of his flow charts for salmon management (Figure 1). It shows how the assumptions inherent in the salmon story and the fragmented institutional structure create a behavioral context that leads to the listing of salmon under the federal ESA. Figure 1 does not show the complete behavioral context for salmon management. A complete treatment would extend back through the cultural filters to the ostensible purpose of salmon management. The statements in the bold-lined boxes were convenient and somewhat arbitrary starting points.

During the course of my career I have had letters written about me by various organizations or individuals representing special interest groups. Those letters called for my dismissal or at least disciplinary action. Once, after I gave a presentation on the problems associated with large scale private salmon ranching in Oregon, I was confronted by a professor of fisheries who asked me, "Why are you against everything we stand for?" By we, I suppose he meant the community of fisheries biologists who he believed stood for fish factories.

When I finished mapping the behavioral context of salmon management, it dawned on me that, throughout my career, all of the biologists I've known who found themselves in hot water did so because they conscientiously did their job. Yet they were considered troublemakers and problem employees. When Dave Bella introduced me to his concept of behavioral context, I immediately understood that those troublemakers had stepped outside the norms of the behavioral context. Their behavior was driven by a different set of assumptions and beliefs, one more in line with the ostensible mission of the agency than the industrial production system. When individuals in an organization step outside the behavioral context they are considered strange, disloyal troublemakers. That sends a powerful message to the whole institution: "strive for normalcy and let the context (story) determine your behavior." And, what if this context has been wrong for a century or more?

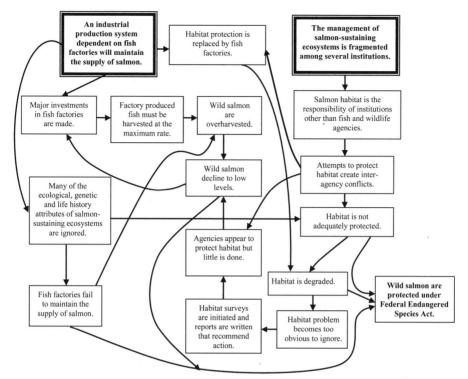

Figure 1: Partial behavioral context for salmon management. Read the statement in any box, then follow the line forward to the next box with the word "therefore" or follow the line backward with the word "because." For example, look at the box near the upper left in the diagram. It says "Factory produced fish must be harvested at the maximum rate." Going forward, therefore "Wild salmon are overharvested" and going backward, because "Major investments in fish factories are made." Two statements (bold-lined boxes) start the flow diagram. One is a summary of the salmon story and its assumption that a simplified production system would maintain the supply of salmon. The other describes the fragmented institutional structure.

At this point we can stop our examination of the hidden mass of the iceberg called the salmon's problem. It's time to swim to the surface, take a deep breath, and think about the previous four chapters. The salmon's problem begins with the cultural filters through which we perceive, engage, and experience the natural world. Do we see commodities and landscapes that must be dominated and exploited or do we see a world that evokes a sense of wonder and a respect for the gift of life that we share with Douglas-firs, cedars, winter wrens, oxalis, and salmon? Those cultural filters with their emphasis on control and exploitation shaped our vision of the salmon's production system and the stories we tell about it. The salmon story and the industrial production of salmon were strengthened by the constraining bureaucratic boundaries in ecosystems fragmented among several government institutions and the special interests they serve.

Where is all this heading? Our sense of place will continue to erode and the powerful and ubiquitous markets will replace the annual return of the salmon as the place-defining event. By the time we notice that the Pacific Northwest has been culturally, ecologically, and physically homogenized by market forces, by the time we notice that the uniqueness of place has been lost, the return of the wild salmon will no longer be the place-defining event—and no one will notice they are gone. Salmon feedlots and engineered habitats will have replaced salmon-sustaining ecosystems. Wild salmon will remain in the Pacific Northwest in the images on T-shirts, coffee cups, and posters—market products. The wild salmon's replacements crowded into feedlots, or the salmon that stay in factories their entire life never tasting the salty waters of the sea, will resemble their ancestors in appearance only. They will have no more of the characteristics that constitute a wild salmon than the images on T-shirts. All this will come to pass unless we rewrite the story. Before we can achieve salmon restoration, we need to achieve *re-story-ation*.

Listed below are the mission statements taken from the websites for Idaho, Washington, and California Departments of Fish and Game.

Idaho
All wildlife, including all wild animals, wild birds, and fish, within the state of Idaho, is hereby declared to be the property of the state of Idaho. It shall be preserved, protected, perpetuated, and managed. It shall be only captured or taken at such times or places, under such conditions, or by such means, or in such manner, as will preserve, protect, and perpetuate such wildlife, and provide for the citizens of this state, and as by law permitted to others, continued supplies of such wildlife for hunting, fishing and trapping.[33]

Washington
To preserve, protect and perpetuate fish, wildlife and ecosystems while providing sustainable fish and wildlife recreational and commercial opportunities.[34]

California
The Mission of the Department of Fish and Game is to manage California's diverse fish, wildlife, and plant resources, and the habitats upon which they depend, for their ecological values and for their use and enjoyment by the public.[35]

Side Channel 4
A Look at the Year 2150

From time to time while writing *Salmon Without Rivers* and this book, I would push back from my desk and shift my thinking away from the immediate problems facing the salmon and speculate about the future. In those speculative moments, I would think about where our current path is going and where it will take the salmon and the bioregion. Sometimes I imagined taking a journey forward a hundred or more years in time and thought about what I might find. This side channel and the next describe two of those trips.

〜〜

I don't think the state can afford to have 40 percent of its water supply go to waste in the ocean, when there are needs for it in other parts of the state.
—William Gianelli, Director of Water Resources, California.[1]

I am going on a most unusual fishing trip. I am taking my great, great, great … grandson fishing. My plan is to take him to the river that has always held a special place in my heart. At first, I was somewhat nervous. Seven generations separate Charlie and me, but there is something about his looks and behavior that I find familiar. Within a few minutes we are at ease with each other and the chance to go fishing makes us instant pals.

Since this is an imaginary trip, I don't have to explain how I happen to have my old pickup truck or how I managed to get the fuel to run it. As we head out of town, I run into the first big change since 2012. All the major highways and most of the secondary thoroughfares are toll roads and the privilege of using them comes at a high price. My pickup dwarfs the few smaller vehicles on the road, and they all appear to be either solar or battery powered. There are no eighteen-wheelers. At a rest stop, one guy looks the pickup over and tells me about a fellow in Portland who has a barn full of old vehicles with combustion engines. "He and his friends drive them in the Rose Parade every year." Then he adds, "He doesn't use gasoline though. I'm not sure what he uses."

Charlie is studying dinosaurs in school, and, as I drive, he tells the same stories about Tyrannosaurus and Brontosaurus that my own boys brought home from school, stories about changing environments and disappearing giants. Charlie's reference to changing environments strikes home particularly hard. I expected to see change on this trip, but the magnitude of the change is staggering.

There is little in the infrastructure that looks familiar. A few roads follow the same old routes, but I recognize little else. The few new or well-maintained buildings are small islands surrounded by aging and poorly maintained structures. Even the parks and other natural areas within the city look worn out, over used, and under maintained. The Columbia and Willamette rivers are about half the size I remember. I wonder what happened to the water. Everything I see suggests several decades of a very poor economy. In spite of the crumbling cityscape, I hold out hope that the river to which I am taking Charlie will still be a special place. I tell myself, "Once we get out of town, things will be different."

I am really anxious to see the river. It was a special part of my life. For my entire life, the river was a friend and a refuge where solitude and contemplation came easy. My sense of place was rooted in and nourished by its flowing waters. I tell Charlie, "There is a river near here, and when I was younger I spent a lot of time there. I learned how to fish for salmon on that river."

I see familiar signs in the landscape that signal we are approaching the river, but the river is not where I think it should be. I tell Charlie, "There are so many changes that I must be confused about where we are." I do a quick check on our location. The distance traveled is right, and I recognize some features in the landscape. The river should be here, but where is it? A strange thought crosses my mind, but I dismiss it. Rivers don't just disappear. You can't lose them like your car keys. Then a sign grabs my attention: Ben Franklin Regional Water Center, next exit. Maybe the folks there can clear up the mystery and give directions.

In the lobby of the main building, a series of displays and a brief film explain the center's purpose and operation. I quickly learn what happened to my river; it and the other rivers in the area no longer exist. They were replaced by "an efficient, engineered water capture and delivery system. Rivers are now considered quaint symbols of an earlier, wasteful period whereas today water is recognized as a precious commodity and it is captured, managed, and distributed by the Regional Water Centers. Wasteful rivers were replaced by a thriving water industry." A man wearing a uniform with a badge that says he works for the center comes by and asks if I have any questions. His nametag says he is a fishery biologist. My head is so full of questions I don't know where to begin.

"What does a biologist do in a place like this?" I ask.

He explains that he manages the fisheries in water storage areas where the existence of fish is compatible with the other uses of the water. He tells me, "In those areas where fish are allowed, my job is to minimize impacts on the fisheries due to the center's operation." Those words are very familiar. They are the same old fishspeak. My blood begins to simmer as I recall biologists one hundred and fifty years ago giving out the same reassuring nonsense: "Don't worry about hatcheries, habitat degradation, or excessive harvest because we are going to minimize the impacts on salmon." It's comforting language and one hundred and fifty years ago I thought such talk was devoid of meaning. I see now that those words may have been devoid of meaning, but they had consequences beyond my wildest imagination. My blood starts to boil as I tell the biologist in a voice as calm and as controlled as I can muster that if the impacts were minimized they would be zero and that is clearly not the case. He says that they are minimizing the impacts. The catch, angler days, output from hatcheries, and licenses sold in this area have remained steady for the last five years.

"What catch?" I ask.

"There are places in the Ben Franklin system where fish are allowed to live, mostly bass, bluegill, and some walleye, and a few other species." I drop the subject, realizing that it is pointless to get angry at this biologist for conditions whose roots go back to my time and beyond. The biologist is riding a shifting baseline, and this is all normal to him. Some things don't change.

At first I am hesitant, but then I ask, "What about salmon? There used to be Chinook salmon in the rivers around here."

He responds, "That was a long time ago. If you want salmon you have to go north—northern British Columbia and Alaska. Unfortunately, salmon need rivers, and we just can't waste water by letting rivers dump their water into the sea like we used to."

The biologist tells me about a special place, one of three in the state, where a person can still fish for trout. "It's clearly identified, just watch for the sign," he says. It's not far, so Charlie and I head for the pickup. At this point I would be happy to find a gravel pit with people standing shoulder to shoulder fishing for catchable trout.

"There it is." Charlie is pointing at the sign that tells us that trout fishing is five miles. I follow the arrow and begin preparing Charlie for his encounter with the wily aristocrat of flowing waters.

The trout stream the biologist told us about is a half-mile stretch of river between two water collection and distribution reservoirs. Fish are stocked immediately above our numbered fishing site as we wade out into the faux stream. It's not really a stream but a concrete channel designed to look like one. The boulders are concrete, and the logs are also imitations. I try to ignore the man planting our allotment of fish and concentrate on Charlie, who is all smiles. Later, at the check station, the cashier hands me a bill for $750 for the ten fish Charlie caught. "That's a bit steep," I say.

The cashier says, "The fishery has to be self-sufficient. The price of the fish must cover all the costs including the wasted water." I wonder what water was wasted, but keep my thoughts to myself. As she makes out the receipt, she tells me about a time when a person could catch all the hatchery fish they wanted and even several salmon for a fee of $40 a year.

As we head toward home, I wonder if this trip was really a good idea. As I reflect on the day's events, I get angry: angry at myself because I did not do more when I could, angry at all the self-serving and short-sighted decisions, and angry at all the complacent fools (me included) who stood by while the seeds of this disaster were being planted.

We are hungry so I pull up to a homey-looking café for a hamburger. The place is empty except for an elderly man at a corner table reading a book. I didn't see a sign out front identifying the café as vegetarian, but it is obvious from the menu. When the waitress comes to take our order, I say we were hoping to get a hamburger. She gives me a curious look and says, "None of the regulars here can afford hamburger, and you don't look like you are carrying around bags of money either. So what'll it be?" We order a couple of soy burgers.

The old man comes over and introduces himself. "I'm curious," he says. "Your truck and your comment to Helen, the waitress, tell me you are not from around here. You appear to not only be from a different place, but, and I hope I'm not being offensive, you seem to have come from a different time."

I laugh it off and say, "Well, at times I feel that way, but you are right: I am not from here." I think to myself, "I am definitely not from here." Then he says he is a retired history professor so I ask him to sit with us a few minutes because I have a few questions about the history of the area. It takes just a little prodding for him to shift into teaching mode, and the story of the past one hundred and fifty years begins unfolding.

It's obvious that he has told this story many times before; he has it down pat.

"Our present predicament has roots that go back a long ways, at least to the nineteenth century and the industrialization of our economy. However, I don't want to bore you with two centuries of history so I'll cut to the chase. A little more than midway through the twenty-first century, we reached a crossroads. Oil supplies were beginning to decline, and it was easy to see the end of cheap energy approaching, and cheap oil was the fuel that drove the whole global economy. As the oil supply began shrinking, our way of life, institutions, economies, the things we had taken for granted began to unravel. It wasn't just a problem for us here in the West. It was a global problem."

He gives me a sly look and says, "I'm sure you remember some of this from your history classes." Then he continues on while Charlie and I dig into our soy burgers.

"What happened at this crossroads has been analyzed by sociologists, anthropologists, ecologists, and just about every kind of ologist there is.

They all agree we made a bad choice and left the crossroads on the wrong path. There was still time; we could have launched a massive research and development effort to restructure the source and use of energy and a parallel effort to restructure the economy from a global-corporate approach to regional economies based largely on local resources. Back in the waning years of the twentieth century people called these regional economies bioregionalism. The really sad footnote to this story is that, beginning in the 1970s, every American president pledged to break the country's dependence on oil. There are a lot of reasons why those efforts failed, but basically the status quo was just too easy and the corporations protecting the status quo were just too powerful.

"As I said, we still had time, but the magnitude of the changes taking place made people apprehensive. Our political leaders not only failed to prepare the country for those changes, but they stoked that apprehension into fear, then intensified and exploited it. Instead of taking the country on a search for alternatives to the global oil-based economy, they fought a series of expensive wars to secure the last few drops of oil. The country took on so much debt that other nations quit lending us money. At that point political leaders compounded their mistakes. To cover the debt run up by their wars, they started selling off the country. We were flying full speed down a dead end road looking back at how great we once were and not forward to where we were headed. To top it off, economists gave the whole thing a positive spin. They predicted a utopian existence once everything was put into private ownership. The roads were first to go. So we started paying tolls to foreign corporations for the privilege of driving on our roads. Then the ports were sold, followed by farms and food-processing infrastructure, our forests, coal, oil (the little we had left), and finally our water.

"The regional water center you passed up the road captures nearly all the water in several catchments and then sells it to the highest bidder. It's my understanding that a big share of it goes to southern California and parts of Arizona."

"You mean the Ben Franklin Facility?" I say.

"Umph! Ben Franklin. Yes, all the water centers are named after our country's founding fathers or other patriots. It's supposed to soften the fact that all the centers and the water are owned by transnational corporations or foreign governments. Well, I've taken up too much of your time. I suppose you need to get this young man and his fish home."

He goes back to the corner table and his book, while Charlie and I head for home.

On the drive back, Charlie is happily telling the story of each trout that he caught. Then he asks, "If you could have any wish, what would it be?"

I say, "I'll have to think about that a little."

"I know what my wish would be," Charlie says. "I wish that I could go fishing on that river you told me about."

As I think about Charlie's question, the events of the day keep crowding into my head. I am reminded of a Jane Siberry song from the 1980s called "Bound by the Beauty." In that song, Siberry tells us that she is coming back in five hundred years and the things she loved better still be here. I think to myself, "Jane, you are going to be disappointed." Then I think about Siberry's song in a different way. What if the people one, two, or five hundred years into the future could bring back the self-serving and shortsighted politicians, the agency administrators with feet of clay, and hold them accountable for their decisions. Then I remember Charlie's question and I say, "You know, Charlie, I guess I would make the same wish. I would like to take you fishing on my river." But Charlie doesn't hear me. He is already asleep.

Part 2—Re-story-ation

Only economists still put the cart before the horse by claiming that the growing turmoil of mankind can be eliminated if prices are right. The truth is that only if our values are right will prices also be right.

—Nicholas Georgescu-Roegen[1]

Put more bluntly, we believe the current pressures on resources are caused directly and indirectly by increasing human numbers through many different avenues ... Our conclusion is that population growth must be designated as the alpha factor.

—Gordon Hartman, Thomas Northcote, and C. Jeff Cederholm[2]

THE PASSAGES BELOW DESCRIBE TWO RIVERS. BOTH WERE ONCE THE HOME of steelhead, and the Carmel still is, but its steelhead population is listed as threatened under the federal ESA. Read the two descriptions, and I'll catch you again on the other side.

The Carmel River

The Carmel is a lovely little river. It isn't very long but in its course it has everything a river should have. It rises in the mountains, and tumbles down awhile, runs through shallows, is dammed to make a lake, spills over the dam, crackles among round boulders, wanders lazily under sycamores, spills into pools where trout live, drops in against the banks where crayfish live. In the winter it becomes a torrent, a mean little fierce river, and in the summer it is a place for children to wade in, for fishermen to wander in. Frogs blink from its banks and the deep ferns grow beside it. Deer and foxes come to drink from it, secretly in the morning and evening, and now and then a mountain lion crouched flat laps its water. The farms of the rich little valley back up to the river and take its water for the orchards and the vegetables. The quail call beside it and the wild doves come whistling in at dusk. Raccoons pace its edges looking for frogs. It's everything a river should be.

—John Steinbeck[3]

The Los Angeles River

From its beginning in the suburbs of the San Fernando Valley to its mouth at the Pacific Ocean, the river's bed and banks are almost entirely concrete. Little water flows in its wide channel most of the year, and nearly all that does is treated sewage and oily street runoff. Chain link fence and barbed wire line the river's fifty-one-mile course. Graffiti mark its concrete banks. Discarded sofas, shopping carts, and trash litter its channel. Weeds that poke through cracks in the pavement are the only plants visible along its course. Fish larger than minnows are rare even where the river does contain water. Feral cats, rats and human transients are the dominant animal life on its shores.

—Blake Gumprecht[4]

Did Steinbeck accurately describe the Carmel River circa 1945? I don't know, but an accurate picture of the river was not the reason I included the passage. I used it as a description of what all rivers are capable of and what they ought to be allowed to be. Clearly Steinbeck believed rivers are the focal point for a community of beings including fishermen, children, and farmers, as well as frogs, mountain lions, deer, and foxes. The river defines a place not only as a physical presence, but as the setting for a series of place-defining events and relationships predicated on clean, flowing water. Steinbeck's rivers have value beyond the commodities that can be extracted, packaged, and sold.

Before proceeding, I need to add a postscript to Steinbeck's description of the Carmel River. The current condition of the Carmel is certainly not what it was in 1945. It suffers from many of the same problems that afflict the Los Angeles River. In 1999, American Rivers placed the Carmel on its Top Ten Most Endangered Rivers list.[5]

The Los Angeles River shows what happens when a river's only value is the monetary worth of its commodities. Los Angeles River water was converted to a commodity of such high worth that no other attributes or values mattered. Water was extracted and sold until the river ceased to exist; its ghost now roams the empty concrete channel. Water is a river's soul. Flowing water sustains the web of relationships that is the essence of a river ecosystem. Unfortunately the value we assign to those relationships is not recognized by the market. The Los Angeles River is an example of what happens when an ecosystem is simplified and reduced to the market value of its commodities, which are then sold to the highest bidder.

In our management of salmon and their watersheds, we carry John Steinbeck's Carmel River in our heads and believe that vision is what we are working towards. However, in actual practice, salmon management is based on a myth embedded in story. At the heart of this myth is the belief that natural, salmon-sustaining ecosystems can be replaced by an industrial production system that serves the market and market values. While the rivers of the Pacific Northwest have not reached the same level of degradation as the Los Angeles River, the shameful state of the salmon suggests that they are moving in that direction. An approach to salmon management guided by a story that only embraces market values has not and will not sustain the regional icon.

An approach to salmon and watershed management that reflects Steinbeck's vision will require rewriting our current salmon story, rebuilding the internal machinery of salmon management institutions, and changing their behavioral contexts. Chapters 5 and 6 describe the steps that I believe will accomplish those goals and put salmon management on the path to recovery. But first I need to deal with two eight-hundred-pound gorillas: human population growth and global climate change—and a few words about the word *inevitable*.

Throughout most of my career, when salmon biologists discussed human population growth, it was tied to the inevitable degradation of freshwater habitats and an inevitable increase in the demand for more sport and commercial harvests. They usually agreed that the inevitable increase in the market's demand for fish and the inevitable habitat degradation would lead to the inevitable need for more fish factories. Those discussions were often infused with a palpable sense of despair. Who could blame the biologists? Locked in their little piece of a fragmented ecosystem, they had no authority to deal with such big societal issues as population growth and climate change, but at the same time, they knew those issues were threatening the fishery resources they were charged with protecting. Inevitable can be a debilitating and depressing word.

Wendell Berry says that the word *inevitable* has important consequences. It can be used as an excuse to avoid dealing with difficult problems. The word *inevitable* attached to problems like population growth and climate change excuses responsible officials from diving into the difficult work involved in the search for solutions. The word *inevitable* suspends thinking because it says there is only one possibility.[6] With regard to salmon management and recovery, *inevitable* does two things: 1) it once again allows those in charge of salmon conservation to play the role of hapless victims of circumstances beyond their control, and 2) it freezes the status quo in place, locks salmon management into the current story, and emphasizes the use of tools and approaches that have failed for over a century. In the end, *inevitable* creates its own self-fulfilling prophecy.

In his book *Goodbye to a River,* John Graves succinctly stated the population problem, "Men increase; country suffers."[7] The negative effects of human population growth on salmon and their ecosystems have been recognized for several decades. It's a problem that will continue to grow; and even though fishery biologists have no authority to deal directly with it they can and should explain its consequences. Recently, scientists dealing with salmon recovery were criticized for failing to speak out forcefully regarding the consequences of human population growth.[8]

Before giving you my take on the population problem, I want to recommend three papers written by fisheries scientists who didn't flinch from addressing human population growth and the problems it creates for wild salmon:

"Science and Management in Sustainable Salmonid Fisheries: The Ball is Not in Our Court."[9]
"Human Numbers—The Alpha Factor Affecting the Future of Wild Salmon."[10]
"Wild Salmon in the 21st Century: Energy, Triage and Choices."[11]

I like these papers because they not only discuss the effects of population growth on salmon, but also suggest remedial actions to mitigate those effects. For example, the second paper lists eight steps that would limit human population growth or its effects on natural ecosystems and salmon. The steps range from a limit on immigration to incentives that would slow population growth. While I agree with the assessments and recommendations in the three papers, I'm going to approach the problem from a different perspective.

Human population growth will certainly strain the resources and spaces we share with salmon and add to the existing obstacles to their survival. But salmon were in decline in the early 1900s when the number of people living in Oregon was less than one quarter of the population in 2000. Does that mean we have to go back to a population size prior to the first signs of the salmon's decline, back to the 1880s? That's not possible. If this is the only solution to the problem of population growth, we can stop thinking and accept the inevitable extinction of wild salmon. But don't give up just yet. Put the word *inevitable* aside for a while and think about the population problem from a different perspective.

That perspective is based on this hypothesis: the *relationship* between humans and salmon may be of equal or greater importance than the actual *number* of people. It was the nature of their relationship that sustained the Native Americans and salmon for at least four thousand years before the arrival of Euro-Americans. It was the nature of the Euro-American relationship with salmon that led to depleted and extinct salmon populations within a hundred years after they settled in the Pacific Northwest. A few people with a destructive relationship with nature can do as much or more harm as a larger population with a healthy one. Our distorted sense of place, the beliefs and myths that underlie the salmon story, and the fragmented governance of ecosystems have created a relationship between us and the wild salmon that has undermined the latter's productivity for over a hundred years. More people will intensify the destructive effects of that relationship, but we can change our relationship with salmon. That is a step we can and must take. The place to begin changing our relationship with salmon is with the government bodies (tribal, federal, state) charged with primary responsibility for salmon conservation—the salmon management agencies.

At this point there are some who will argue that changing the human-salmon relationship will be as difficult as reducing the number of people. I disagree. For at least fifty years since the publication of Rachel Carson's *Silent Spring*, the people of this country have been searching for and slowly developing a better relationship with nature. At a broad, national scale, progress has been slow and, at times, punctuated with reactionary periods—the George W. Bush administration, for example. When the scale is narrowed to the conservation of wild salmon, there are also signs of progress. The rapid growth in the number of watershed councils reflects the public's growing concern for the survival of local salmon populations. Some environmental organizations are finally using the courts to challenge the operation of fish factories and reduce their effects on wild salmon. Changing the salmon story will require a lot more work, but it's not an impossible task.

By suggesting that we focus on the human-salmon relationship, I am not dismissing the importance of population size, but I am also not surrendering to the *inevitable* loss of salmon. Historian Arthur McEvoy tells us that giving up our attempts to dominate nature and building a different relationship with it "need not entail the exchange ... of reason for animism or industrialism for [a] hunting-gathering economy. It only means learning to care for other living things, which is at once a special talent and the special responsibility of the species."[12]

The other eight-hundred-pound gorilla is climate change. It's a global problem with global origins, although some countries contribute proportionally more to the problem than others. Because climate is highly variable and because the cause of climate change cannot be pinpointed to a single activity or country, it provides the unscrupulous or uninformed with opportunities for obfuscation, denial, and obstruction. I hope President Barack Obama takes the United States out of the ranks of the obfuscators and adopts a meaningful leadership role on this problem. Even if the world, with the leadership of the United States, were to get serious about doing something about the problem, climate change is already here, and its effects are being felt today. The predicted effects in the Pacific Northwest do not bode well for the salmon. To deal with climate change we will have to give up our belief that a simple, industrial production system can replace the accumulated wisdom of sixty million years of salmon evolution.

Salmon have a long history of surviving major changes in climate, as well as catastrophic changes in the landscape. They survived the advance of the Wisconsin Ice, a period of very warm climate eight thousand years ago, volcanic eruptions, massive lava flows, mountain building, and gigantic floods. If salmon can survive all that, why would climate change pose a bigger problem? It's a serious problem because we have been systematically stripping the salmon of their evolutionary legacy; stripping them of their genetic and life history diversity; stripping the rivers of the natural infrastructure that creates healthy and complex habitats. Through our misguided story, we have been breaking the links in the life history-habitat chain the salmon must follow to survive. In a real sense, we are burning the salmon survival candle at both ends. On one end, our salmon story is eroding the attributes (biodiversity) the salmon need to adapt to climate change, while at the other, by not seriously addressing the causes of climate change, we are creating the conditions that that will require all of the salmon's adaptive capacity to survive. Individuals can and do have a responsibility to reduce their use of fossil fuels, but the major changes needed to arrest the full impact of climate change will have to come from programs of national and global scope. However, we don't have to wait until the world's political leaders muster the will to act to help the salmon survive climate change. We can focus our attention on the other end of the survival candle and stop the practices that are reducing the salmon's inherent ability to adapt to changing conditions. This also means changing our salmon story.

These two eight-hundred-pound gorillas pose serious threats to the continued existence of Pacific salmon. The current salmon story cannot deal with those problems and, in fact, it is making them worse. In spite of these formidable obstacles, I am not ready to give up on a substantial recovery of wild salmon. My steadfast belief that the recovery of wild salmon is a real possibility comes from what I have learned from these magnificent animals and the people of the Pacific Northwest who still hold them in high regard. However, two reasons for my position stand out: evidence that people, when informed of the issues, have decided to protect salmon even at substantial economic cost, and what I have learned from the salmon themselves about their potential to recover in response to real stewardship.

Sometimes people underestimate society's commitment to salmon conservation. In the 1950s and '60s, when the program to build large mainstem dams in the Columbia River was under way, there was a similar proposal to develop the hydroelectric potential of British Columbia's Fraser River. The biggest threat to the Fraser's salmon was the proposed 720-foot-high Moran Dam—a mainstem dam located about two hundred miles upstream. But, unlike the Columbia River dams, Moran was not built. The people of British Columbia choose to protect their salmon runs and forgo the economic benefits that Moran would have created. At the same time, the people in the United States chose to put their salmon at risk.

In the Columbia River, in spite of skepticism by some biologists, the public was told that hatcheries would mitigate the effects of the dams—a position consistent with the salmon story. The message given to the citizens of the Pacific Northwest was reduced to a slogan: "salmon and power, we can have both."[13] Fish factories were traded for the loss of mainstem habitats. In the Fraser River, biologists told the public it had to choose: high dams or salmon—but it could not have both. Canadian biologists examined the record of hatchery mitigation then dismissed it as unworkable. The citizens of British Columbia chose salmon over the economic benefits that the dams would produce.[14]

I haven't given up on recovery, because the wild salmon themselves give me all the reason I need to work for their recovery. Recently one of the three sockeye salmon populations in the Columbia River demonstrated their ability to respond positively to effective stewardship. This story is about the Okanagan River/ Osoyoos Lake run from the upper basin in British Columbia. The Columbia River sockeye, like the basin's other salmon, have a long history of declining abundance. The commercial fishery for sockeye was closed in 1972, but runs remained depressed for the next thirty-five years.[15]

In 2008, the sockeye experienced a dramatic change. From an average run of about seventy-two thousand fish for the previous thirty-five years, the count at Bonneville Dam jumped to 213,602 and then the increased returns continued—177,823 in 2009, 386,525 in 2010, and 185,796 in 2011. The run for 2012 is projected to surpass the larger run of 2010. Most (about 80 percent) of the sockeye are returning to the Okanagan/Osoyoos River and lake system and most of those are wild. A hatchery program contributes about 10 percent to the total return.[16] What caused the big increase in wild salmon even though their spawning grounds are above nine dams on the Columbia River?

Some factors outside the Okanagan/Osoyoos system—such as improved ocean conditions and improved passage survival at the mainstem dams in the Columbia River—contributed to the increase. But three changes in management practices also played an important role. The sockeye escapement targets had been capped below the ecosystem's carrying capacity. In 1999 sockeye escapement targets for the Okanagan/Osoyoos were revised upward allowing more wild fish into the spawning areas. Additionally, a decision-support system helped managers control water flows reducing density independent mortality on eggs and fry. The same decision-support tool was used to reduce oxygen depletion and temperature constraints on juvenile rearing habitat in Osoyoos Lake. These actions boosted the production of sockeye smolts from an average of three hundred thousand per year to an average of two million with a high of nine million smolts.[17] Clearly the managers were repairing links in the life history-habitat chain. The salmon were responding to stewardship based on the ecology of the fish.

This is just one example of salmon, steelhead, and trout responding positively to management practices that emphasize ecological relationships rather than commodity production. In so many ways, wild salmon have been telling us that we need to strengthen the ecological ties between salmon, people, and place. But somewhere along the way we lost our ability to listen, to listen to the land and rivers and to listen to the salmon. They are trying to tell us that depletion, degradation, extinction, and impoverishment are not inevitable.

The next two chapters describe steps we can take to put our relationship with salmon on a sustainable path. Chapter 5 addresses the low-hanging fruit, the obvious steps that are relatively easy to implement. Agency administrators already have the authority to take many of the actions recommended. All that is really needed is the courage to break away from the status quo, but it will also take pressure from the public and political leaders.

Chapter 6 returns to an idea that I introduced in *Salmon Without Rivers*, but didn't fully develop. In that book, I introduced the readers to Tom Jay, who is one of the few individuals I know with a strong spiritual connection to the salmon. Tom says that salmon recovery will take nothing short of remaking our culture. In *Salmon Without Rivers*, I said I did not know how to begin discussing such a fundamental change. At the time, I was thinking about cultural change in an either/or frame. I thought the current culture needed to be completely replaced by one that was more salmon friendly. Then I read Bryan Norton's book *Sustainability: A Philosophy of Sustainable Ecosystem Management*. Norton argues for a pluralistic approach to the resolution of environmental problems, an approach that legitimizes several different points of view and values in the debates leading to a shared vision of the future. In Chapter 6, I return to Tom Jay's plea for cultural change and describe four ideas that should be part of Norton's pluralistic mix.

Chapter 5—Beyond the Crossroads:
First Steps Toward Salmon Recovery

All attempts to manage are attempts to tell a story about how the land ought to be, and by definition all these stories are simpler than the world itself.

— Nancy Langston[1]

For dangers are being produced by industry, externalized by economies, individualized by the legal system, legitimized by natural sciences and made to appear harmless by politics.

— Ulrich Beck[2]

Someone once asked me: Why is it so hard to change fish factories and reduce the problems they create for wild salmon? The real problem is not the fish factories, it is the story that guides how hatcheries are used. The story creates the demand for fish factories, elevates their importance, and excuses their failures. That is the key message from the first four chapters of this book, and incorporating that message into your understanding of the salmon's problem is the first step beyond the crossroads.

During the last twenty-five years of my career, I committed a lot of time to the study of the assumptions and beliefs that underlie salmon management, but I didn't discover their importance and their relationship to the salmon's status on my own. For that, I am indebted to two remarkable fisheries scientists. Over several months, Dr. Charles Warren[3] and his colleague Dr. William Liss and I met regularly over lunch to discuss natural resource management. Quite often we talked about the importance of the story or, to use their terminology, the conceptual foundation. At the time, they were professors in the Department of Fisheries and Wildlife at Oregon State University and I was the assistant chief of fisheries at the Oregon Department of Fish and Wildlife (ODFW). Charles and Bill approached the discussions from a theoretical perspective, while I wanted to know how those theories could be applied to salmon management. I brought several questions to the meetings—unanswered questions that I had acquired during my tenure at ODFW. For example: With so many individuals within the department concerned about the fate of wild salmon, why do activities detrimental to the wild salmon persist? Why didn't agency staff take full advantage of the opportunity to protect wild salmon when the first wild fish policy was adopted in 1978? Why were high harvest rates maintained in the face of evidence that escapement of wild coho salmon was declining to dangerously low levels? In spite of growing evidence of problems, why did fish culturists continue to ignore the effects of their operations on wild salmon? Why was habitat protection given rhetorical rather than real emphasis?

Through their patient explanations, Charles and Bill helped me understand the source of salmon management's detrimental actions and why they diverge from the agencies' laudable goals. The meetings took place long before I read about John Livingston's environmental iceberg. They prepared me to see the value in that metaphor and helped me understand the story's pivotal, but hidden, role in salmon management. Charles and Bill motivated me to first study and understand the existing story and then to try to change it. Looking back on that time I can see that I was pretty naïve about how easy it would be to make those changes. I tried to take a quiet, low-key approach using probing questions about specific activities or programs. Quite often the response was a puzzled look. Some responses were defensive—"You are not being loyal to the agency" or "Why are you trying to expose our dirty laundry?" My favorite was being called un-American and lacking in patriotism because I kept asking questions about the performance of fish factories and the need to hold them accountable to improve their performance. Hatcheries apparently rank right up there with mom and apple pie as sacred symbols of America.

A few years after leaving ODFW, I was appointed to the Independent Science Group, later renamed the Independent Scientific Advisory Board, a panel of eleven senior scientists and managers charged with examining the scientific basis for the massive salmon restoration program in the Columbia River. Bill

Liss was already on the panel and, while the whole panel supported an analysis of the program's conceptual foundation, three members—Jack Stanford, Rick Williams, and Phil Mundy—besides Bill and I were strong advocates for such a study. Chip McConnaha was the coordinator for the Independent Science Group and, while not a member of the group, he supported our examination of the conceptual foundation for the Fish and Wildlife Program.

It took several years of work by the panel before the results were published in the book, *Return to the River: Restoring Salmon to the Columbia River.*[4] Two chapters in the book were devoted to the conceptual foundation. The book and a summary of it published in *Fisheries* magazine concluded that the conceptual foundation of the salmon recovery program in the Columbia Basin was flawed and did not reflect the latest science. The advisory board developed an alternative consistent with the existing science.[5] The initial response to the book was something like: "You guys just don't understand salmon management." Later it shifted to: "This is nothing new. We adopted your proposed conceptual foundation a long time ago." They may have adopted the appropriate vocabulary, but there was little evidence in their programs that they had adopted the real substance of a new story. Through experience I learned how difficult it is to change salmon management's underlying story. It's not a simple narrative that can be rewritten or edited into a different form. The story is not like a broken part in a machine that can be pulled out and replaced with a new one.

My experience convinced me that changing the salmon story requires an indirect approach. Accordingly, this chapter describes a series of steps, which will, if conscientiously implemented, confront the failings of the status quo and expose the weakness in its underlying myths. The goal is to make attachment to the status quo more uncomfortable than the search for a scientifically credible alternative.

From time to time in my career, I encountered biologists who seemed to have little interest in protecting habitat, but loved to rebuild degraded or destroyed streams, especially if the work involved the use of heavy machinery. There also were biologists who believed that hatcheries were the solution to most management problems. Those individuals will not find the following recommendations very satisfying. These recommendations are aimed directly at the salmon management agencies. However, I believe it would be beneficial if educators at the university level, especially in fisheries programs, discussed these concepts with their students. The university classroom is where students and professors can explore alternative ways of thinking about salmon and place.

The fate of the salmon is not inevitable. We have the intellectual and technological power and the moral obligation to coexist with these magnificent animals, to coexist in a nurturing relationship. I hope we have the resolve to use that power to the benefit of salmon, people, and place.

Hold Salmon Managers and Administrators Accountable

It's my view that salmon recovery has not failed because of a lack of scientific information, funding or legal mandates. It has failed because responsible elected officials and governmental agencies at all levels—provincial, state and federal—have not used sound ecological principles when engaging in salmon recovery efforts. Additionally, they have not been held accountable for doing so by society at large.

—Jeffery Dose[6]

Before beginning this section, I looked in the index of several recent books on fisheries management, including my previous book. I looked for the words *accountability* or *accountable* and failed to find them. I am sure that within the thousands of pages in those books some of the authors do mention the need for accountability, but it is clear that it is not given a prominent place in narratives about fisheries management. I believe the lack of real accountability is directly related to the miserable state of the world's fisheries. Callum Roberts succinctly stated why salmon managers avoid accountability: "Experience has a bitter taste in fisheries management."[7]

What does it mean to say that fisheries management and recovery programs need accountability? For example, should agency administrators be held accountable for the often-used promise to "double the runs of salmon"? Accountability would certainly cause them to pause and take a closer look at the technical basis for their programs before they made another doubling promise. A commercial fisherman once suggested that the administrators of fish and wildlife agencies should have the size of their salaries tied to the size of the salmon runs. While I'm sure this form of accountability would be popular with some, it's not what I have in mind.

About ninety years ago, biologists in the Pacific Northwest recognized that, in spite of their best efforts, salmon abundance was declining. This motivated them to write the first recovery plans. Recovery planning has been a part of salmon management ever since, and, given the number of recovery plans produced, one could conclude that the Pacific salmon are the most restored species group in the world. Obviously, most of those recovery plans failed to accomplish their goals. As part of the research for *Salmon Without Rivers* and this book I have read dozens of recovery plans including a few from the 1920s and '30s. I learned from that experience that salmon recovery and management programs have had a disturbing lack of historical perspective.[8] While the authors of those plans may have mentioned the history of salmon decline, the documents were largely ahistorical, disconnected from previous attempts to protect or to recover Pacific salmon abundance.

I want to be clear what I mean when I say that there is an important relationship between accountability and a healthy historical perspective. From a historical perspective, accountability means that the decision makers and authors of recovery plans today must acquire a working knowledge of earlier efforts—what was done, and why it succeeded or failed. A failure to understand the history of salmon recovery efforts ensures a continuous recycling of failed approaches and the persistence of the status quo.

The experience of reading so many past and present recovery plans led to three questions that I ask of each new plan I read:

Did the authors demonstrate that they understand what actions were proposed in the earlier recovery plans for the same or similar salmon populations and watersheds?

Did the authors explain why the earlier recovery actions either succeeded or failed?

Did the authors describe how their plan will avoid the mistakes and failures of the past?

Any salmon management or recovery plan written today that does not clearly answer those questions should not be approved until it does. But achieving this level of accountability requires pressure from a vigilant public and political leaders.

Some management and recovery plans fail because they are simply not implemented or are implemented halfheartedly. This is especially true where the plan tries to break with the status quo. The organizational structure of an agency is geared to implement the status quo, so it's unlikely that a plan that deviates from the norm will fit into the established operational machinery of the agency. Such a plan will wander the halls of the bureaucracy like an orphan looking for a home, until it finally comes to rest on the top shelf of a seldom-visited back room. I ask two additional questions of any plan I read to determine if it has a real chance of being implemented, or if it will be exiled to the back room. Those questions are:

Do the authors show that the agency has an organization structure capable of implementing all the elements of the plan?

Do the authors describe who is responsible for carrying out all tasks and how their workloads would be shifted to accommodate the new duties?

The inertia generated by the status quo will overcome the best plans unless the operational machinery of the organization is changed to complement the new work.

An example illustrates the importance of accountability from a historical perspective. In the 1930s, the plans to build mainstem dams on the Columbia River were ramping up. Dams had long been recognized as a threat to the persistence of salmon. To mitigate this threat, salmon managers were asking for more fish factories. However, there was enough concern over the ability of hatcheries to compensate for the effects of the dams that the Secretary of the Interior appointed an independent Board of Consultants to evaluate the mitigation plans. The board recognized that there was substantial uncertainty surrounding the effectiveness of mitigation hatcheries so they recommended that they be treated as an experiment and cautioned salmon managers that there was the possibility of failure. They added: "The adoption of the plan for trial should not be understood as implying an indefinite commitment to its support, but only for so long as the results may reasonably appear to justify its continuance."[9] This was a reasonable approach given the persistent failure of fish factories. Time passed, dams and fish factories were built, and the salmon runs declined to the point that they needed protection under the federal ESA. Unlike the Columbia River, stilled behind mainstem dams, fish factories, in spite of their failure as a mitigation tool, continued to roll on.

Sixty years after the Board of Consultants tagged the mitigation plan as an experiment, a growing concern prompted the Northwest Power Planning Council to undertake a review of hatchery mitigation in the Columbia River.

The result was a set of policies regarding the use of hatcheries in the basin and a plan to reform hatchery operations. One of those policies struck a familiar note: "Artificial propagation remains experimental. Adaptive management practices that evaluate benefits and addresses scientific uncertainties are critical."[10]

Because the council didn't recognize the earlier failure to implement an experimental approach, it didn't take the steps needed to ensure the experiment was actually carried out. It was about this time that a new use for hatcheries called supplementation was gaining momentum. The council's policy should have ensured that supplementation hatcheries would be treated as experiments. However, in its 2003 review of supplementation, the Independent Scientific Advisory Board concluded that supplementation was being implemented in a way that would make comprehensive evaluation unlikely.[11] And so it goes: historical failures are ignored and then repeated. Accountability begins when salmon managers are not allowed to repeat the mistakes of the past.

Of course, there are several other questions that must be asked of new salmon management and recovery plans. Many of those questions deal with how the relevant fishery science is used or how the problem the plan is supposed to overcome is defined. Those questions are the basis of most technical reviews. The five questions that I have linked to accountability are rarely, in my experience, a part of technical reviews, but they should be. The state of the salmon is such that agency administrators must not be allowed to retreat to the comfort of the status quo and continue to recycle its failures.

Accountability requires more than recognizing the long history of salmon management and restoration programs and absorbing what they can teach us. As this chapter progresses, accountability will resurface in the discussion of other steps on the path to a new salmon story.

Put Institutional Learning on the Agency Agenda

We are an action-based society. We are a nation of doers. … it's easy to see that, if you can get people to lead with their actions instead of their thoughts, they will be forced to rationalize their thoughts and emotions into line with their actions.

—Charles Hayes[12]

In 1997, I wrote a paper that included the following statement:[13]

> *Institutional learning involves more than well-designed studies and the accumulation of knowledge by individual scientists. It is the ability to incorporate the collective wisdom of resource managers and scientists into agency programs, policies and guiding philosophies. Since it is the institutions that persist and provide continuity, rather than individuals, it is at the institutional level that relevant learning takes place. However, there are many impediments to institutional learning[14] including an emphasis on legitimizing established programs and practices and a reluctance to legitimize institutional evaluation.[15]*

Institutional learning and accountability are closely linked. The lack of accountability lets salmon management institutions avoid learning. If an institution's senior administrators are not held accountable, there is no need to learn from past mistakes and successes. Conversely, an institution that lacks a formal learning process cannot be held accountable. Without learning, any attempt to enforce accountability is met with meaningless spin and positive slogans. Given the deplorable status of Pacific salmon, the lack of institutional learning is hard to understand and justify.

To avoid confusion, it's important to differentiate individual and institutional learning. All the public and private organizations that I worked in or around had several intelligent individuals who were genuine experts in their field of specialization. In aggregate, they were a rich source of information, and this pool of talented individuals is an important attribute in any organization. However, those talented individuals retire or move to new jobs in other agencies. When they leave, their knowledge goes with them. Too often, this store of knowledge held by individuals is not incorporated into the institution's memory.[16]

An example will illustrate the difference between individual and institutional learning. Several years ago, I attended a meeting hosted by the Washington Department of Fisheries to discuss the depressed status of and possible recovery actions for pink salmon in the Dungeness River on the Olympic Peninsula. The Dungeness had two populations of pink salmon, one in the upper river and one in the lower. At the beginning of the meeting a geneticist working for the agency came in and made an excellent presentation on the biology and genetics of the two populations. The bottom line was this: the two populations used different habitats and they were genetically different so it would be important to avoid mixing them when attempting to increase their abundance. The geneticist left the room after his presentation. An administrator spoke next. He started with a statement the gist of which was: OK, now it's time to be realistic and practical. He went on to propose a simple approach to the problem. Wild fish will be collected from only one of the populations and taken into a hatchery and spawned. The resulting juveniles will be planted into both populations. Pink salmon from the upper and lower river populations would be mixed. The geneticist had provided important information based on the latest science, but the institution failed to incorporate that knowledge into its programs and policies. It failed to learn. This incident was just one of many I have encountered that demonstrated, as Dave Montgomery wrote, "Despite rhetoric to the contrary, modern management of salmon and their habitat provide a superb example of maladapted management—the failure to learn by experience."[17]

Institutional learning occurs when administrators and senior biologists look for new information, evaluate its relevance, and, when appropriate, incorporate it into the agency's programs and policies. This process cannot be treated as an additional duty that is addressed only when there is extra time. It must be a priority and administrators must actively participate. Finally, institutional learning cannot be a distortion of real learning where agency officials simply sort through new information, cherry picking those parts that support current programs and policies.

Kai Lee describes two modes of institutional learning: single loop or double loop. In single loop learning, an institution tries to solve all its problems using the existing story.[18] That is what the administrator in the example was doing. He was solving the problem of diminished pink salmon runs using an approach consistent with the old salmon story even though that approach was not supported by information held by the individual experts in the agency. Single loop learning helps create and reinforce the behavioral context of an institution.

Writing on the use of single loop learning by the military, Lieutenant Colonel John Nagl described it this way:

Most people so restrict their frame of reference, or context, for the problem they are facing that little change can occur. They get into such a routine with their work that they view virtually all problems in a similar way—back to all problems looking like nails when all you have is a hammer.

Consequently, when asked to change matters, they tend to operate in a confined "single loop" of learning on which they can only do "more of" or "less of" the same thing because of the given context.[19]

When an institution's story fails to solve problems, Lee says it must switch to double loop learning. The institution must challenge the status quo and revise its story to make it a useful problem-solving tool. But Lee recognizes the difficulty involved in double loop learning and concludes: "in many cases double loop learning is not even attempted."[20] For a long time, I have felt that my university education would have been enhanced by an exposure to the concepts described by Lee.

Conceptual foundations (the story) should always be considered a work in progress. They should be continually updated and revised as new information and scientific understanding of the salmon-sustaining ecosystems emerge.[21] Double loop learning should not be a rare event, but a routine part of an institution's operation. However, because it challenges the status quo, individuals who see the need for double loop learning in a fish management agency often find themselves outside the accepted behavioral context of the institution and become isolated and subjected to cold indifference and uncertainty regarding continued employment.

While I was at ODFW, most of the attempts at double loop learning took place within the department's Research Division. This should not be surprising. Research biologists work on the cutting edge of fishery science. But work at the cutting edge may simply focus on new ways to implement the status quo, and that is generally what occupied researchers at ODFW prior to the 1970s. Research focused on how to raise more and healthier juvenile salmon in hatcheries. That work was important and led to needed improvements in fish culture. But in the 1970s a gradual shift took place.

Researchers began evaluating the performance of hatcheries. A wild fish policy was adopted in 1978, which, while not effectively implemented, encouraged research into the effects of hatcheries on wild salmon. By the 1990s this work was coming into direct conflict with the existing salmon story. The Research Division was becoming the source of double loop challenges to the status quo. As Lee predicted, the agency did not appreciate or see its value and responded by disbanding the Research Division, so that it was no longer an autonomous unit within ODFW. The researchers were put directly under the control of regional managers "to make them more responsive to management," which is code for making sure research supports the status quo instead of challenging it. A few individuals whose work constituted a serious challenge to the status quo found that the funding for their positions was suddenly threatened. The message to staff was clear: stepping outside the behavioral context and challenging current programs and policies can be dangerous to the health of your career.

Ray Hilborn wrote a paper in 1992 with the provocative title "Can Fisheries Agencies Learn from Experience?" In that paper, Hilborn emphasizes the need for adequate monitoring and evaluation as a prerequisite to institutional learning.[22] However, consistent with the lack of institutional learning, inadequate monitoring and evaluation of salmon management and recovery programs is a chronic problem. For example, the Independent Scientific Review Panel (ISRP) evaluates the scientific quality of projects proposed for funding through the Northwest Power Planning Council's massive Fish and Wildlife Program in the Columbia Basin. In addition to the review of individual project proposals, the ISRP periodically writes retrospective reports that summarize the results of their work. In their 2007 retrospective, the ISRP confirmed the chronic lack of adequate monitoring: "Perhaps the most dominant recurrent theme through all the ISRP's past retrospective reports was the lack of adequate monitoring and evaluation in fish and wildlife projects throughout the fish and wildlife program."[23]

Several times during my career, I have heard someone justify the lack of monitoring and evaluation by a statement similar to this: We don't need to waste money on monitoring and evaluation. We should put all our funds into projects that get things done. The pervasive nature of this attitude is why I put the quote from Charles Hayes at the beginning of this section. A long time ago we rushed into the implementation of an industrial salmon production system, and since then we have been trying to justify that decision while avoiding any real evaluation of it. The chronic lack of monitoring and evaluation is able to persist because no one with the authority has bothered to ask: How are we doing? The lack of adequate monitoring and evaluation persists where there is no formal learning process and especially where there is no accountability.

Hilborn's paper is worth reading for all the information it contains, but for this discussion, I want to call attention to one additional point: the idea that a robust institutional memory is a prerequisite to institutional learning. Information that does not become part of an institution's memory has no chance of being incorporated into policies and programs. Clearly, if an institution is to learn, it has to have a mechanism that captures and retains information. Hilborn said information is stored in files, reports, computer databases, and people's minds. But it is not enough to simply store information; it has to be passed on from one generation to the next; it has to become a part of the agency's cultural milieu. An agency can facilitate the transfer and use of information in six ways: short experience reports (relating what was learned in implementation of a policy, program, or experiment), old timers' seminars, division historians, an annual review of what was learned during the year, memoirs of retiring staff, and new staff training.[24] All of these are useful tools in the quest for institutional learning. In addition, the main point raised in the previous section (authors of new programs must demonstrate that they have studied and learned from the past) would also enhance institutional learning.

The incorporation of new scientific understanding into salmon management and recovery programs is slow to nonexistent and the lack of institutional learning is the major culprit.[25] Effective institutional learning needs strong leadership that will make it a high priority. It also requires a leader who will encourage rather than punish those who take on the difficult task of challenging the status quo through double loop learning.

Stop the Practice of Shifting Baselines

Frameworks that maximize the use of fisheries history would help us to understand and to overcome—in part at least—the shifting baseline syndrome.

—Daniel Pauly[26]

Those of us who follow the status of Columbia River salmon are often treated to feats of magic that even the best practitioners of the art would envy. This magic occurs about every eight to ten years with the appearance of a "record" run of salmon; 2010 was one of those years. With a projected run of 470,000 spring Chinook, *The Oregonian* reported that it "could be the largest spring Chinook run on record in the Columbia River."[27] Well, not quite. When the Northwest Power Planning and Conservation Council estimated the losses of salmon due to the operation of the federal hydropower system, it started with an extensive analysis of the historical salmon abundance. Their estimates of the average predevelopment run of spring Chinook ranged from 710,000 to 1,530,000 fish. But that was the average run size, and *The Oregonian* was talking about a "record run." The council also estimated historical run sizes based on maximum peak harvest, and those estimates can serve as surrogates for past record runs. For spring Chinook, the historical record run ranged from 1.7 to 2.3 million fish.[28] A return of 470,000 fish was an improvement over the number of spring Chinook in recent years, but it was hardly a record. This is an example of a shifting baseline.

Marine biologist Callum Roberts described the problem of shifting baselines this way:

> *Early accounts of the abundance of fish and wildlife offer us a window to the past that helps reveal the magnitude of subsequent declines. They provide us with benchmarks against which we can compare the condition of today's seas. Such benchmarks are valuable in countering the phenomena of shifting environmental baselines whereby each generation comes to view the environment into which it is born as natural, or normal. Shifting baselines cause a collective societal amnesia in which gradual deterioration of the environment and depletion of wildlife populations pass almost unnoticed.*
>
> *Whatever the reason, if people forget what the seas were once like, and consider today's waters as something approaching natural, then we could end up trying to maintain marine ecosystems in their present degraded states. We have to do better than that.[29]*

The same could be said about rivers in the Pacific Northwest and their salmon. When fish and wildlife institutions employ a shifting baseline, they are acknowledging the failure to achieve their mission to protect salmon, hiding the magnitude of the loss and their failure. Shifting baselines are an impediment to institutional learning and they help administrators of fish and wildlife agencies evade accountability. Because a shifting baseline conceals the loss and even creates "record" runs out of impoverished abundance, administrators believe there is really no need to search for an alternative approach to management or to learn from mistakes.

Setting the historical abundance of salmon as the baseline does not mean that I am advocating a return to the world of 1800. That is not, and should not be, the goal of salmon management and recovery programs. But we should always be aware of the real loss that has taken place and where our recovery efforts lie relative to the historical condition. For obvious reasons managers and administrators do not like historical baselines. They make them confront the real status of the salmon and management's failures. Here are some of the feeble excuses managers and administrators use to avoid confronting their agency's performance: "Setting the historical abundance of salmon and the historical conditions of ecosystems as the baseline is like driving down the road only looking in the rear view mirror." I agree that such driving would be problematic, but if you don't look in the rearview mirror once in a while you will never see the flashing red or blue lights or other signs of danger, and maybe that's a part of the salmon's problem too. Some administrators will tell you when confronted with the historical record, "Well, yes, things are bad, but they would have been a lot worse without our efforts." There is some truth in that statement, but not as much as they would have you believe. If they really believe that that statement is a real measure of success, then the mission of the institution should be changed to: Don't let things get quite as bad as they could.

Oregon's Independent Multidisciplinary Science Team summed up the importance of maintaining historical baselines in salmon recovery programs with this statement:

> The historic range of ecological conditions in the Pacific Northwest,
> both habitat and of salmon stocks, is important because it provides a
> framework for developing policy and management plans for the future. The
> performance of salmonids under historic ecological conditions is evidence
> that these habitats were compatible with salmon reproduction and survival.
> Land uses resulting in non-historical ecological conditions may support
> productive salmonid populations, but the evidence for recovery of salmonids
> under these circumstances is neither extensive nor compelling.[30]

It's understandable how the administrators of salmon management agencies can be lured into the trap of shifting baselines. They provide an escape from the record of long-term decline. Today's administrators of fish and wildlife agencies should not be blamed for the massive impoverishment of salmon that occurred before their time, but they should not be allowed to conceal the problem by shifting the baselines forward, just as they should not be allowed to keep repeating the same management approaches that have failed to protect the salmon.

Management agencies should use the best available information and reasonable assumptions to reconstruct the historical abundance and distribution of all the salmon populations they manage. That historical information should then be part of any reporting the agency does regarding projected run sizes, recovery targets, escapement goals, and habitat restoration. Any time an agency biologist or administrator discusses a current or projected run size they should juxtapose that number with its historical antecedent. Here is a hypothetical example: A river had a historical run of eighty thousand Chinook salmon, whereas the current run hovers around a thousand fish. The management agency prepares a recovery plan that is designed to double the run to two thousand fish. Instead of simply stating, "We are going to double the run," which sounds like a big deal, the agency administrator would have to state, "We are going to double the current run and, if we are successful, the size of the run would be 2.5 percent of the Chinook salmon's historical abundance in this river."

Charles Darwin's work left us with many important lessons. One of those lessons was this: history matters. It's not just the long, evolutionary history that matters, but also the shorter-term history of our relationship to the place where we live and the animals we share that place with. Shifting baselines are an attempt to deny that history and its relevance.

Hatcheries Need a New Metaphor

Stocking catchables [trout] worked, as one fish and game commissioner from the state of Connecticut put it in 1938, if you "keep it under control and face the facts by deliberately treating it as a manufacturing proposition."
—Anders Halverson[31]

Metaphors? Aren't metaphors typically considered a symbolic use of language, a rhetorical tool, not an important contributor to thought or action? Perhaps you are wondering, what role do metaphors play in the serious business of salmon recovery? Actually, they play an important role. According to linguists George Lakoff and Mark Johnson, our conceptual frameworks and stories are largely metaphorical; and a major purpose of this book has been to show that those conceptual frameworks influence the thoughts and actions that define our relationship with the natural world. According to Lakoff and Johnson, "The essence of metaphor is understanding and experiencing one kind of thing in terms of another."[32] In Chapter 1, I pointed out how the machine metaphor for nature exerts a strong influence over the way we understand and experience ecosystems: fish factories are one of the consequences. Early fish culturists—as indicated in the quote that opened this section—equated hatchery practices to a manufacturing process thereby acknowledging their link to the machine metaphor for ecosystems. They actually called the early hatcheries fish factories.[33] Modern fish culturists try to soften the hatchery's image by embedding it in a pastoral setting of manicured lawns and tidy ponds full of little salmon gobbling up processed food tossed onto the water. But, in fact, they are factories in the larger industrial food production system. Using the factory as a metaphor for hatcheries not only reflects what they were actually called in the nineteenth century, but how we have understood and used them ever since.[34]

When we understand and experience hatcheries as factories it focuses attention on factory-related things such as production efficiency, product output, facility capacities, food conversion rates, and other concerns that are common to most factory operations. Hatchery records are replete with factory-like statistics that give information on operational efficiency, the acquisition and use of raw materials (fish eggs, feed, and water flow) needed to make the product, and data that track the growth of product to the target release size. Those records stop at the hatchery fence. They rarely contain information on the effects of fish factories on wild salmon once the factory-produced fish are released into a river. The lack of information on the ecological consequences of hatcheries reflects the long-standing attitude that their operation is independent of the ecosystems where their products are released. The factory metaphor for hatcheries focuses attention on processes and problems within the hatchery fence and renders concerns about the ecosystem and wild salmon irrelevant.[35]

Factories achieve economic efficiency by making products that are uniformly similar. Fish factories do the same, and their focus on an economically efficient operation renders salmon biodiversity an impediment to an efficient factory-like operation (see Chapter 2). Fish factories are operated as though salmon life histories are invariant—as though there is only one life-history pathway. Biologist Robert Bugert summed it up this way:

Few hatcheries can flexibly promote varied life histories, primarily because of rigid propagation strategies. Not only do supplementation programs disregard the realities of variable environmental conditions, but they also relentlessly battle ever-changing stream flows, debris loads, and other vagaries of nature. Heraclitus declared "you cannot step in the same river twice." Monolithic hatchery structures do not perform well in protean stream systems.[36]

Many of the newer state and tribal hatchery programs—and a few older ones such as Oregon's Elk River Hatchery—are beginning to look beyond the hatchery fence, but as the earlier chapters demonstrated, past beliefs and practices still plague salmon management. For example, historically, the two performance measures used to evaluate fish factories were the number of eggs incubated or the number of fry or smolts produced. Those indicators measured performance of the fish factory's acquisition of raw material (eggs) and output (smolts). If managers were really paying attention to what happens after the factory-produced fish are released, the performance criteria would have changed from the number of juveniles released to the contribution those fish actually make to the sport and commercial fisheries and the escapement to the river. That change, as obvious as it is, has not taken place. In its review of a 1999 performance audit of Oregon's coastal and Willamette River hatchery programs, Oregon's Independent Multidisciplinary Science Team made this observation:

. . . 41 of the 51 salmon and steelhead programs reviewed in the audit used smolt releases as a performance measure; 34 of the 51 salmon and steelhead programs used egg takes as a performance measure; and only 9 of the 51 salmon and steelhead programs evaluated performance using adult returns. These data clearly show the emphasis on activities inside the hatchery—egg takes and rearing juveniles to the smolt stage. Survival to adult, which takes place outside the hatchery, received little attention.[37]

For more than a century, salmon managers have tried and failed to replace the historical productivity of natural, salmon-sustaining ecosystems with the products from fish factories. The hatcheries were bound to fail, because salmon managers ignored the ecological processes and relationships that sustain salmon. What we have been doing is more akin to the proverbial "bull in a china shop" than a reasoned attempt to integrate fish culture into the ecological processes of natural, salmon-sustaining ecosystems. Fish factories have been run on the cheap. The fish and wildlife agencies have tried to get by without investing in the research needed to understand the ecosystems they were trying to enhance, nor have they invested in the monitoring and evaluation needed to understand fish culture's consequences. I believe correcting this deficiency should start with a new metaphor for hatcheries, a metaphor that helps us experience and understand hatcheries from an ecological perspective.

There are probably several metaphors that would place hatcheries in an ecological context; I'm going to propose using the tributary stream. Think of hatcheries as tributaries (albeit mechanical ones) embedded within a watershed; tributaries that are home to a population of artificially propagated salmon. When hatcheries are understood as tributaries, when they are understood as part of a salmon-sustaining ecosystem, they can no longer be considered independent of the ecosystem where they are located. Several questions will be raised that currently fail to surface in the operation of fish factories.[38] For example, does the hatchery's output replicate production from a small or large tributary in the watershed? Is the level of production compatible with the capacity of the river and estuary and does the scale of the hatchery's production pose competition or predation risks to the wild population? Do the environmental conditions in the hatchery mimic the natural conditions one would expect in a tributary to the watershed? Is the hatchery's rearing and release schedule consistent with the life histories and life history diversity of wild populations in the watershed? Will the hatchery create a mixed-stock fishery and, if it will, is the fishery managed to protect the wild salmon in the watershed?

Treating hatcheries as though they are tributaries changes the role and responsibility of the hatchery manager. Managers of "tributary" hatcheries need a solid grounding in the ecology of the watershed's natural and artificial production systems. In other words, to effectively integrate natural and artificial propagation, the person making decisions about how the hatchery is used should be the best salmon ecologist in the watershed.

Asking someone to do something different is much easier than asking them to think about what they are doing in a different way. Changing the metaphor will take strong leadership—leadership that is persistent in asking and demanding answers to questions that arise from a different, more ecological metaphor. And it's not necessary to totally abandon the machine metaphor of ecosystems and the factory metaphor for hatcheries. They are useful for resolving some problems, but they shouldn't be the only tools a salmon manager has. A metaphor that better represents the ecology of salmon-sustaining ecosystems will loosen the dominance of the machine over our thinking and reduce its negative side effects. Thinking of salmon and their ecosystems in a different way is an important step in changing the salmon story.

Undertake Hatchery Reform

Over the past thirty years, there have been several attempts at hatchery reform. They made little difference. Having watched as the different attempts at reform fell flat, I am convinced that they will not be successful until hatcheries are viewed through the lens of a different metaphor, until salmon managers think about what they are doing in a different way. In the meantime, recent hatchery reform projects have produced an extensive catalogue of recommendations that will be available when the salmon managers are receptive to change.

The latest attempt at hatchery reform is being conducted by a panel of experts called the Hatchery Scientific Review Group (HSRG). The HSRG has produced several important and useful documents on the subject. Among all those documents is one sentence that may be the most important sentence regarding hatcheries that has appeared in the fisheries literature in several decades. The sentence is, "Hatcheries are by their very nature a compromise—a balancing of benefits and risks to the target populations, other populations, and the natural and human environment they affect."[39] This, in my opinion, is a groundbreaking statement. Stating in unequivocal terms that hatcheries are a compromise eliminates the myth that they are benign enhancement tools. It says that the benefits that hatcheries produce always come at a cost. It forces the salmon manager who wants to use the fish factory tool to explain what the compromise entails for each hatchery program. That explanation must state in explicit terms the tradeoffs, the cost to wild salmon, the manager is willing to accept to obtain the anticipated hatchery benefits. Until the HSRG's unequivocal statement, management agencies were able to easily deflect concerns about the effects of hatcheries by claiming that there was uncertainty as to whether there were any negative effects or by simply stating that they would minimize any negative effects.

Before continuing I need to discuss the use of the word *minimize* and its frequent use or misuse in salmon management and recovery documents. The word has been used to avoid explaining in plain, understandable English the compromises associated with fish factories. Salmon managers have so misused *minimize* that it has lost its meaning. They often say they are going to minimize the effects of a hatchery on wild salmon populations or minimize habitat degradation. That should mean the effects will be so small as to be negligible. But, generally, that is not what they are telling you. They are using the word in a relative sense. They are really saying that the negative effects of a hatchery's operation will be consistent with the tradeoffs—the risks to wild salmon— they are willing to make. They don't want to tell you what those tradeoffs are because often they don't know themselves. Those tradeoffs can radically change what *to minimize* actually means. So every time the word is used, it should be followed by a description of the tradeoffs that are acceptable, and a description of the monitoring program that will determine if minimize as defined is actually achieved. The meaningless way it is now used has inflicted a lot of damage on wild salmon populations.

Because the ecological processes in natural, salmon-sustaining ecosystems were ignored as our dependence on fish factories grew, it is not likely that salmon managers will have the information needed to explain the ecological tradeoffs that their hatchery programs require. So the first step in hatchery reform is to develop budgets for each hatchery that include the cost of a monitoring and evaluation program designed to collect the information needed to understand the hatchery's ecological consequences. These additional costs should be borne by the hatchery programs and not taken out of habitat protection and restoration or other management activities. In the meantime, the ecological costs for each hatchery should be assessed using information currently in the literature, applying it to specific hatcheries with conservative assumptions and analysis. This should include, but not be limited to, the following steps:

Quantify risks to wild populations for all hatcheries. Describe the acceptable tradeoffs in natural production and the monitoring program that will ensure the tradeoffs will not be exceeded.

The impacts of hatchery programs should be reported as life stage survival rates of affected wild populations.

Each year determine the cost to produce a harvested fish from each hatchery program and provide that information in a form accessible to the public.

Adopt a stock transfer policy that prohibits moving fish and eggs between watersheds.

Use the latest scientific information regarding hatchery impacts on wild salmon to develop a set of standards for hatchery operations. Those hatcheries that cannot meet the standards within three to five years should be closed. The standards should be peer reviewed before implementation.

Return to Place-based Salmon Management

Wisdom sits in places ...

—Keith Basso[40]

Wild Pacific salmon are place-based animals. Their migration covers an immense area including river, estuary, and ocean. Mature, wild Pacific salmon return to the same stream reach where they began life as a fertilized egg to close the cycle and create the next generation of salmon. As I discussed in Chapter 1, the salmon's attachment to place is the source of their biological diversity and the wellspring of much that make the salmon and steelhead treasures deserving of real stewardship.

Native Americans were also bound to specific places. They were enmeshed in a web of relationships that defined a community of native plants and animals, all of which had equal standing. Over time, the exchange of gifts among members of this community strengthened those relationships. Place became more than geographical space; it evolved into the sacred home of the tribe.[41] Throughout the Pacific Northwest, those sacred places almost always included the home of one or more wild salmon runs. It's tempting to say that Native Americans managed their local stock of salmon, but the common use of the word *manage*—control and manipulate salmon to achieve specific ends—does not describe what they did. It's more accurate to say the Native American culture participated in coevolved and sustainable relationships among salmon, people, and place.

With the arrival of Euro-Americans, the relationships among salmon, people, and place weakened and over time practically disappeared. Euro-Americans ignored the salmon's attachment to place. The fisheries moved off shore and targeted groups of salmon made up of fish from several different rivers. Fisheries targeting mixed stocks of salmon couldn't ensure that enough fish from each of the populations escaped the fishery to seed their spawning areas. Unlike the

Native Americans' harvest of their local run of salmon, the Euro-American mixed-stock fisheries were disconnected from any consideration of place. They overharvested the smaller and weaker stocks and, as they declined and disappeared, no one paid much attention (See the quote from W. F. Thompson near the end of Chapter 1).

Eliminating or weakening the relationship between salmon and place made it possible to concoct a salmon story based on an industrial production system. Hatcheries are not tied to place. They can be located anywhere. The eggs they incubate may be from fish that were headed to distant spawning grounds. The juvenile salmon may be out-planted to streams also a long way from their ancestral spawning and rearing areas. The rearing regime in the hatchery is often very different from the salmon's natural life histories, life histories tuned to the biophysical qualities and attributes of their ancestral stream. Fish factories produce animals with simple life histories adapted to concrete ponds of uniform design, rather than the unique and varying habitats of their home streams. The result is that salmon factories make products whose real purpose is to support markets. Salmon management that doesn't recognize the salmon's strong connection to place is a weak defender of habitat, and it is headed for failure.

In Chapter 1, I discussed Gary Nabhan's idea that the loss of relationships is an important cause of extinction. When an industrial salmon production system is superimposed onto a watershed, it weakens or eliminates the salmon's relationship to place and starts the slide towards extinction. The recovery of wild salmon cannot proceed until we recognize and restore their strong relationship to place and create a management story that reflects the importance of place. That means: vigorous protection of habitat and the diverse life history pathways that extend out from the spawning areas; setting ecologically meaningful escapement targets and funding adequate monitoring of those targets; setting harvest regulations that reflect the priority of achieving adequate escapements back to all spawning areas; stopping the transfer of salmon from their home streams to foreign waters; and demanding that hatcheries integrate their operations into the watershed's natural, salmon-sustaining ecosystem.

There is reason for optimism. The rapid growth of watershed councils following the listing of salmon under the federal ESA is a positive step toward place-based management. Watershed councils are groups of citizens that organize for the purpose of restoring their local stream and its salmon run. The public, if not the managers, appear to be reconsidering the importance of place. Citizens are trying to mend the relationships to their local salmon population and its home stream and in the process they are achieving a renewed sense of place. There are 159 watershed councils in Oregon alone.[42]

Citizens of the Pacific Northwest are not satisfied with the prospect of their salmon surviving only in fish factories and feedlots or in faraway places like western Alaska. They want to rekindle the intimate relationships between people and their local salmon runs. Watershed councils are an important resource that could be mobilized into a grassroots campaign to change the salmon story and bring about a better approach to salmon management. So far such mobilization and organization of the individual councils doesn't seem to be encouraged or facilitated[43] by, for example, the Oregon Watershed Enhancement Board.

Watershed councils usually receive guidance from salmon biologists working for state or federal agencies. They need to be careful that such guidance, embedded in the old story, doesn't have them repeating past failures.

Paying attention to ecological relationships and mapping life history pathways onto watersheds requires paying attention to the details and attributes of place. It means designing management programs that are compatible with the evolutionary history of wild salmon. Placeless management may be compatible with Henry Ford's assembly line, but it is antithetical to the salmon's evolutionary history. Placeless management will inevitably lead to more homogenization and simplification of ecosystems and greater industrialization of ecological processes, including the food webs of both humans and salmon. Wild salmon are place-based animals, and their sustained recovery requires a reinvigorated sense of place and a management paradigm that recognizes the importance of place.

Here are several essential steps that will begin reconnecting salmon management to place:

Set an escapement target for each breeding population and establish a program to monitor compliance. Base the initial escapement target on the number of eggs needed for full seeding of habitat then, over a period of not more than six years, increase the escapement target to achieve an ecological standard based on the need for nutrient enrichment.

Manage harvest to achieve the escapement targets described above.

Harvest management involves two primary activities: setting the allowable harvest and allocating it among the different sport and commercial fisheries. The former is a technical task; the latter is a political task. Too often the technical and the political tasks are so intertwined that their separate roles become confused. To avoid that confusion the two activities should be clearly separated (with separate oversight and supervision paths) within the organization structure of the management agencies.

Develop a catalogue of the genetic and life history diversity of each wild breeding population and periodically evaluate and report to the public on the status of those attributes.

Establish habitat protection and improvement criteria that effectively sustain life history diversity, abundance, productivity, and distribution of native wild fish in each watershed. Pay particular attention to flow, temperature, and structural quality of the habitat.

Fish and wildlife agencies should not avoid habitat protection with this excuse: habitat protection is out of our hands because we do not have the authority to control the activities of other agencies that influence salmon habitat. In such situations the salmon managers should use their bully pulpit as the recognized experts to inform the public of habitat-degrading actions.

Treat Salmon Management as One Long Experiment

Because human understanding of nature is imperfect, human interactions with nature should be experimental.

—Kai Lee[44]

The mechanistic view of the world and its dominant metaphor, the machine, have been useful in solving some problems and advancing technology in many fields. However, when ecosystems are understood as machines, they lose their most important attribute—their evolutionary history and its coevolved relationships. Machines are made to operate the same way over time so history in an evolutionary sense is irrelevant. This gives rise to the idea that rivers with depleted salmon populations are equivalent to an engine with a dead battery and that they can be "jump started" by injecting non-native salmon populations into them.

Understanding ecosystems as machines also leads some biologists to seek stability in salmon production. Machines are capable of maintaining a stable rate of output of products. The same stability is expected from fish factories, but is rarely achieved and then only for short periods of time. Year-to-year variation is a natural occurrence in salmon-producing ecosystems, introducing uncertainty that managers must take into account when setting catch and escapement targets. It's a vexing problem that managers, who equate ecosystems to machines, would like to eliminate. To fall back on an old cliché, I wish I had a dollar for every time I heard an advocate for fish factories say, "We need hatcheries to stabilize salmon production." If the salmon never left the hatchery this might be a valid objective. However, salmon must survive in the river and ocean for a year or more after release from the hatchery. During that time they are beyond the hatchery's protective environment. They must deal with threats to survival that their hatchery experience does not prepare them for. The idea that hatcheries will stabilize salmon abundance is a fantasy that has made its way into real management programs through the medium of the machine metaphor.

When a machine has a problem, you fix it: change a part, change the spark plugs, rebuild the carburetor, replace the water or fuel pump, and so on. When the broken part is replaced, it goes back to running the same as it did before the problem developed. When an ecosystem is understood as a machine, a problem such as the collapse of a salmon population triggers a search for the broken part or the limiting factor, which leads ultimately to a jump start, augmented flows, barging juvenile salmon past the dams, or other mechanical fixes. A common thread running through all of the mechanical fixes is the idea of a linear process with a definite end point. Once the part is replaced, or in the case of barging juvenile salmon, the problem is bypassed, the machine goes back to running as it did before. Using the machine metaphor to solve the salmon's problem has proven to be a fool's errand.

Now here is the really big problem that emerges when ecosystems are understood as machines: operating the machine is understood as simply a matter of repetitive tasks or the occasional need to fix a broken part. I remember a biologist who told me that we will one day have enough information on salmon and their ecosystems to make complete control and accurate predictions commonplace. That kind of thinking, based on the machine metaphor, makes it impossible to think of and deal with salmon-sustaining ecosystems as they really are—an evolving web of relationships that is continuously adapting to the changing biophysical attributes of the landscape.

When a river is modified by a dam or some other human activity, it's not useful to equate the change to a broken or modified part in a machine, because that leads to the belief that someone can replace the part or pull a few levers on the ecosystem machine and correct the problem. Dams change the web of relationships that comprise the ecosystem and, in effect, they shift that ecosystem onto a different evolutionary trajectory. The new trajectory is determined by the quantitative and qualitative changes in the web of relationships. In the Columbia River, mainstem dams pushed the ecosystem off the trajectory it had been on since the end of the Wisconsin Ice Age. The Columbia's current trajectory is more favorable to the production of non-native walleye, bass, and shad and less favorable to the production of Pacific salmon. Thinking of salmon restoration as changing an ecosystem's trajectory, changing the ecosystem's web of relationships, makes it a more formidable, but a more realistic task than simply "jump starting" a dead battery.

Ecosystems are moving targets. They are moving, evolving, and coevolving along a developmental trajectory, the course of which is determined in part by evolutionary history and in part by the way human activities change the web of relationships that comprise the ecosystem. Unlike understanding the operation of a machine, there will be no point at which we can say we have learned all there is to learn about a salmon-sustaining ecosystem. There will be no point at which variability and surprise will be eliminated. Learning, experimentation, and adaptive management must be continuous activities in salmon management. Salmon management is one long, never-ending experiment. Management decisions and policies must be treated as experiments with all the rigor that the word experiment implies.

Define What Successful Salmon Recovery Looks Like

Consequently, the biggest challenge facing the region is not the biological uncertainty associated with salmon recovery efforts, but whether the policy makers are willing to face the difficult task of significantly changing the status quo. Restoration of fish and wildlife in the Columbia River Basin will require difficult decisions and will continue to test whether the region's policy makers and elected officials can find the political will necessary to endorse and implement a scientifically sound salmon recovery program.

—Williams et al.[45]

I said in the book's introduction that government agencies in the Pacific Northwest are spending a billion dollars a year on activities related to the recovery of Pacific salmon. What will that much recovery funding buy us? What will it look like when we have successfully recovered the salmon? Ask a dozen individuals across the Pacific Northwest and you will get a dozen different answers. The salmon are such a powerful symbol that people are willing to approve spending substantial amounts of money on the idea of recovery, without a clear understanding of what they are actually buying.

What would the Pacific Northwest look like if salmon recovery programs were to restore salmon abundance at 40, 50, or even 75 percent of the historical levels? Achieving that level of success will not only require changing the developmental trajectories of salmon-sustaining ecosystems, it would require changing the trajectories and expectations of many human economic activities. Defining successful salmon recovery and how the costs of that recovery will be distributed is an undertaking that requires political skill, vision, and courage. Political leaders will have to dissolve the bureaucratic boundaries that fragment the management of ecosystems. Dissolving those boundaries and distributing the cost of salmon recovery shouldn't be a task that fish and wildlife agencies take on by themselves. It needs strong leadership by elected officials.

John Kitzhaber recognized this problem and actually tried to tackle it during his first term as Oregon's governor. He challenged the political leadership of the Pacific Northwest to take a broader view of salmon recovery—a view that transcends the bureaucratic boundaries. In a speech in 1997 he said:

Finally, we must identify the economic stakeholders, acknowledge and validate their concerns, and make the costs and politics of recovering salmon explicit. ... We must address the concerns of the economic stakeholders in any proposed plan and the costs of doing so must be clearly identified and incorporated as a part of implementation. ... [T]he bottom line is that effective salmon recovery, based on sound science, is going to cost everybody something. What we haven't done so far is figure out who or how much. Until we do that, we cannot really make an accountable decision.[46]

And then in 2000, he said:

In the end, the answer will be a political one—informed by good science—
but based on a set of values and on the degree of economic and ecological
risks the region is willing to accept. It is time to shoulder our responsibilities
and develop a blueprint for action. ... To do so, we must engage the citizens
of the Northwest. Engage them in making clear what is at stake in the
Columbia Basin and what our goal must be in response to this challenge.
Engaging them concerning the alternatives we have to achieve that goal.
Engaging them in describing the trade-offs inherent in each option. And that
will require an unprecedented level of political leadership and collaboration
throughout the region.[47]

Governor Kitzhaber challenged the political leadership of the region to define its vision of successful salmon recovery and how the costs of attaining that vision will be distributed. He called for a vision that transcended the bureaucratic boundaries of a fragmented management system. His challenge called for an unprecedented level of leadership and, unfortunately, that level of leadership is still unprecedented. The massive salmon recovery effort continues, but there is still no common vision. We still do not know what successful salmon recovery looks like or how the costs will be distributed. Governor Kitzhaber laid down a challenge to his fellow political leaders in the Pacific Northwest and apparently their response was silence.

As governor, Kitzhaber practiced what he preached. He initiated a statewide salmon recovery plan that engaged all the state agencies, all the economic interests, in a common goal of salmon recovery. It was an unprecedented step in the right direction, but it didn't go far enough. The goal of the program was left vague and maybe that was necessary to get the cooperation of all the diverse interests. But it also provided a big loophole, a loophole that ensured the persistence of the status quo. The Independent Multidisciplinary Science Team (IMST) was established to evaluate the scientific basis of the Oregon Salmon Plan and it recognized this problem. In a 1999 report, the IMST stated: "In spite of the fact that recovery of depressed stocks is the primary goal of the Oregon Plan and a legal mandate of the Pacific Fishery Management Council, the IMST found no explicit statement of the definition of recovery."[48]

Wild salmon penetrate watersheds of the Pacific Northwest so thoroughly that they come into contact with many, if not most, of the economic activities in the region. Successful salmon recovery must be defined in a way that transcends the bureaucratic boundaries that fragment the management of ecosystems and presents a vision where salmon and economic activities coexist. Until the political leadership in the region accepts this challenge, the efforts to recover salmon will be impeded and the status quo will channel recovery efforts into activities that have a long history of failure.

Unless there is accountability in the development of restoration programs that: recognizes the history of failure in salmon restoration programs; stops using shifting baselines; ensures that institutions place a high priority on learning and using what they learn; recognizes ecosystems are evolving and changing and that management and recovery are one long experiment; transcends the bureaucratic boundaries of our fragmented management structure; those programs will continue to be merely symbolic.[49] Those programs will sound noble, even reasonable, and they will give the illusion that government is doing something, while avoiding what is needed to achieve real progress.

What does successful salmon recovery look like and how will its costs be distributed? Now that John Kitzhaber has been reelected as Oregon's governor for a third term, I hope he uses the bully pulpit of his office to once again take up that challenge.

Create Salmon Refuges

To some people, creating a marine reserve [or salmon refuge] is an admission of failure. According to their logic, reserves should not be necessary if we have done our work properly in managing the uses we make of the sea. Many fisheries managers are still wedded to the idea that one day their models will work and politicians will listen to their advice. Just give the approach time, and success will be theirs. How much time have we got?

—Callum Roberts[50]

In 1991, Willa Nehlsen, Jack Williams, and I wrote the paper "Pacific Salmon at a Crossroads" to call attention to the status of salmon populations throughout the Pacific Northwest. Many salmon populations had declined to the point that their very existence was threatened. The salmon were at a crossroads. The paper captured the attention of Northwesterners. It was discussed on radio talk shows, in newspapers, and in meetings and symposia. All this attention raised an important question: Would the bleak description of the state of the salmon alter the status quo? The subsequent listing of Pacific salmon from several rivers across the Pacific Northwest heightened the public's concern over the salmon's fate and unleashed a river of money for salmon recovery. At the time, I was convinced that this combination of events would lead to a paradigm shift and a new approach to salmon management. I was wrong. So far the additional funding and public concern over the fate of the salmon have failed to make much of a difference. Salmon management and recovery programs have largely stayed on the same path that led to the crossroads, but today implementation of the status quo is better funded.

The state of the salmon that Willa, Jack, and I described was not the first time since the arrival of Euro-Americans that salmon were at a crossroads. In the early decades of the twentieth century two long-migrating species groups—Pacific salmon and North American waterfowl—were in significant decline. Habitat destruction and overharvest depleted both and brought them to a crossroads. Similar problems may have brought salmon and waterfowl to the crossroads, but decisions made by their respective management biologists placed the two groups on very different paths. Waterfowl managers recognized the critical importance of habitat and focused efforts on the protection and restoration of wetlands. Working with political leaders, the waterfowl managers created refuges to protect habitat at key places along migratory routes and in breeding areas. Salmon managers opted for a different approach. They decided to maintain their reliance on fish factories to reverse the decline in salmon abundance. They stuck with the status quo in spite of its record of failure. Because they meshed well with the political and economic approaches to resource management, hatcheries were the tool that managers turned to time and time again in their attempt to halt the declining abundance. That does not mean that there weren't other voices calling for the use of different tools. The idea that entire rivers or parts of them should be protected as salmon refuges has been around for as long as there have been salmon management institutions.

In 1892, President Benjamin Harrison declared Afognak Island, Alaska, a salmon refuge. Unfortunately, in 1892, fish factories were considered an essential part of a refuge. As with most early hatcheries, within a few years poor hatchery practices depleted the salmon run, and the refuge status of the island was forgotten. In the 1930s the buildings were converted to a naval recreation facility. Also in 1892, Commander J. J. Brice surveyed military and other government reservations in the Pacific Northwest in order to find future hatchery sites. He recommended setting the Klamath River Basin aside as a salmon preserve. His recommendation was not implemented.

In 1911, Henry Ward addressed the annual meeting of the American Fisheries Society and reminded his audience that refuges and parks were being set aside for mammals and birds, and said that fishes deserved similar treatment. He recommended that river reaches and even entire watersheds be set aside for the protection of native fish.

In 1928, Oregon voters were asked to consider an initiative petition that would have created fish sanctuaries in all or parts of the Rogue, McKenzie, Deschutes, and Umpqua rivers. The petition failed to garner enough votes.

In 1932, the Washington State Supervisor of Fisheries called for the creation of permanent fish sanctuaries to protect the commercial and sport fisheries.

In 1938, the Oregon State Planning Board recommended that the feasibility of setting aside some rivers as salmon sanctuaries be studied. The recommendation was not implemented.

In 1945, the Lower Columbia River Fisheries Development Program was created to compensate for the massive program of dam construction. One provision in the program called for setting aside all the tributaries below McNary Dam as salmon refuges. This provision needed ratification by the legislatures of Oregon and Washington. The Washington legislature approved the plan, but Oregon failed to ratify it.

The Federation of Fly Fishers published the latest call for salmon refuges in 2004 in their magazine *Flyfisher*.[51]

Calls for the creation of salmon refuges couldn't overcome the "almost idolatrous faith" in hatcheries.[52] This book could be described as a long argument in favor of revisiting the decision made at earlier crossroads, acknowledging that mistakes were made, and taking a different path to move beyond the crossroads.

A major obstacle to making that shift is the fact that our relationship with the salmon is largely economic. Salmon are managed as a commodity.[53] Management of this commodity is more consistent with economic dogma than ecological principles. The economic dogma to which I am referring is the belief in perpetual economic growth. As long as our management of salmon is based on economic rather than ecological principles, the dogma of perpetual growth will eat away at habitat while also directly consuming more and more salmon, leaving behind depleted populations in impoverished rivers. From an ecological perspective perpetual growth is an oxymoron. To avoid this characterization of their dogma, economists and salmon managers rely on the concept of substitutability.[54] The assumption of substitutability goes something like this: if wild Pacific salmon are overharvested or their habitat degraded due to the pressures of perpetual economic growth, don't curtail that growth, just substitute factory-produced salmon for the diminished numbers of wild salmon.[55] When managers accepted the industrial production of salmon, they also bought into the economist's belief in perpetual economic growth and its assumption of substitutability. We arrived at the sorry state of the salmon today because we mistook commodity production for conservation and the management of fish factories for stewardship.

The decision at that earlier crossroads to opt for fish factories instead of salmon refuges was an ecological mistake. It was never recognized as a mistake because the management of salmon is based on the rules of economics rather than ecology. If we are to live in a world where wild salmon persist in healthy ecosystems, humans and their institutions must learn how to coexist with wild salmon in the region's watersheds. However, as long as fish factories are used as a substitute for wild salmon, as long as our relationship with the salmon is based on the rules of economics, our ability to coexist with these magnificent fish will be stymied. We need to reconstruct the human-salmon relationship as an ecological one.[56] Until we do that, the only sure way to protect wild salmon will be in refuges. Over the past couple of decades I've thought a lot about the salmon's problem and its resolution and I cannot envision a path to recovery that does not include a system of salmon refuges.

Refuges may do several things: conserve species for future generations, provide long-term research and educational programs, provide opportunities to hunt, fish, and observe native fauna and flora; some refuges may grow crops for resident wildlife and for profit. But at their core refuges have one most important function. They are created to protect the relationship between a species and place, a place that has all the biophysical attributes the species needs to survive. The term *species* is inclusive, pertaining to both plants and animals.

A salmon refuge is a recognition of the importance of place in the salmon's life history. Not just the place where salmon return to spawn but the whole chain of places that the salmon require to complete their life cycle. Because that chain of habitats may extend for thousands of miles, salmon refuges would be limited to those parts of the chain most vulnerable to degradation—those parts that occur in watersheds. A more specific definition of a salmon refuge contains three key elements:

> It must contain habitats compatible with salmon spawning, rearing and migration, and the biophysical process to create and maintain those habitats over time.
>
> It must contain native populations of Pacific salmon, which are capable of expressing a major part of their life history diversity.
>
> It must be managed to ensure that salmon populations and essential habitats are maintained in the future, and will receive greater protection than is provided by existing laws and regulations.[57]

Side channel 5 gives a description of a hypothetical salmon refuge.

In my experience, the idea of a salmon refuge is often met with skepticism, apprehension, and a level of concern that borders on fear. As Callum Roberts said in the quote at the start of this section, managers view the need for refuges as an admission of failure. Talk of creating refuges evokes apprehension and concern because it is a direct challenge to the basic tenets of our salmon story. To accept the need for a refuge, one has to acknowledge the existence of a strong ecological relationship between salmon and place. That relationship is what our industrial production system was supposed to replace.

The word refuge usually evokes the image of the refuges administered by the U. S. Fish and Wildlife Service: Malheur and William L. Finley National Wildlife Refuges are examples. Oregon has seventeen national wildlife refuges. However, a national wildlife refuge is not the only form that effective refuges may take. A refuge that fulfills the core purpose of protecting the relationship between a species and place may take several different forms. It may surprise you when I say the cultivation of the Pinot noir grape in the northern Willamette Valley was aided by something akin to a refuge that protected the special relationship between the Pinot grape and place. The Pinot grape has been called the "grape of place."[58]

Early in the experimental growing of Pinot noir grapes in Oregon, the mostly young vintners recognized the climate and fertile land in Yamhill County had the attributes of place that Pinot grapes required. At the time, Oregon State's Comprehensive Land Use Planning Maps were being drawn by counties and the state. The vintners worked with county planning staff to designate those lands suited for growing these grapes as prime agricultural lands. In effect, the vintners worked to establish a sort of refuge for the cultivation of Pinot grapes. Oregon's award-winning Pinot noir wine is the result of the recognition that place and a relationship to place matter and sometimes that relationship has to be protected by a refuge. The Pinot story also shows that refuges can take many different forms.

The French have a term, *terroir,* that captures the special relationship between land and the characteristics of what it produces. Roughly translated, it means an expression of place or from the land. Salmon are an expression of place. Sport and commercial fishermen, tribal members, educators, environmental groups, and political leaders must recognize the salmon's strong relationship to place and work to ensure that management agencies protect that relationship by all available means, including refuges.

Even though the task of setting up a system of salmon refuges has many obstacles, there are signs of hope. Conservation organizations such as Pacific Rivers and the Wild Salmon Center are working on ideas that are converging on the concept of refuges. They carry names such as salmon reserves, salmon strongholds, or anchor habitats, but their intent is to produce the same outcome as a refuge. Ultimately salmon refuges are an interim step in the longer-term project of developing a sustainable, ecological relationship between humans and salmon, a relationship that would allow the salmon to flourish in the rivers of the Pacific Northwest.

The steps outlined in this chapter are a beginning, the first steps toward a different approach to the management and conservation of wild Pacific salmon. All but two of the steps can be implemented by the salmon management agencies. The exceptions are creating salmon refuges and defining what success looks like; the agencies would have to work with governors and legislatures to complete those tasks.

In this chapter I focused on the salmon management institutions where the shift to a different story must begin. While reforming salmon management institutions and their approach to conservation are critical first steps toward salmon recovery, they are not sufficient. This effort requires an active, vigilant, and informed public. The public must become engaged in the process of reform and hold the management institutions and the political leaders who oversee those institutions accountable. In asking the public to become engaged in changing the salmon story, I am not asking them to take on an entirely new task. They have in the past applied pressure and helped bring about changes in salmon management. The removal of fish traps and wheels in the Columbia River, adoption of wild fish policies, and changes in fishing regulations are some of the changes the public has help implement.

To prepare for the challenge of changing the status quo, of loosening the comfortable ties to the old salmon story, citizens concerned about the fate of wild salmon need to educate themselves about the salmon's problems and, perhaps even more important, develop a new perspective—a new way of thinking about salmon and their relationships to people and place. The next chapter describes some aspects of this new perspective. The ideas discussed in Chapter 6 helped me see rivers, salmon, and the Pacific Northwest bioregion in a different light. They helped strengthen my sense of place.

Side Channel 5
Another Look at the Year 2150

[W]hen the accumulation of perceived failures significantly exceeds the perceived utility of management, the legitimacy and conceptual coherence of that management institution are weakened to the point where they are vulnerable to challenge and open to fundamental change.

—Christopher Finlayson and Bonnie McCay[1]

I have a recurring vision of an imagined future time in the Pacific Northwest. It always starts with a most unusual fishing trip. The year is 2150 and I am taking my great, great, great … grandson fishing again. My plan this time is to take him to Tillamook Bay to fish for salmon and renew my connection to the bay and its five tributary rivers. I am anxious to see how this important part of the Oregon coast might have changed. We say goodbye to his mom and dad, but before heading out of Portland the two of us take a few minutes to walk around the neighborhood.

I expected to see change, but the magnitude of the change takes my breath away. Urban development is intense, with many high-rise apartment buildings. It appears Portland continued limiting outward sprawl, which the city and others in the state had started to do through Oregon's groundbreaking approach to land use planning adopted in the 1970s. The intensity and density of the development is softened by many parks and green areas. The green areas are obviously used by people in the neighborhoods. In addition to the parks, there are many communal vegetable gardens. Tomato, corn, peppers, beans, squash, carrots, and other vegetables are also being grown in pots and raised garden beds. There are about as many vegetables as flowers and shrubs scattered throughout the neighborhoods and parks. Small shops, cafes, and professional offices occupy the lower parts of the tall buildings. I learn later that many of the residents live and work in the same buildings. People are strolling, stopping to talk, picnicking, enjoying the green spaces, and tending their vegetable gardens. While watching the comings and goings of the neighborhood I recall the writings of Jane Jacobs and think this was her vision.[2] These are real neighborhoods. They are places where a real community of people live and work and, I would guess, look out for each other. In my time, the places where people worked and slept were separate places connected by a tangled web of commuter routes from an office or factory to a house with a side trip to the mall. The whole system was so fragmented it kept people isolated with little chance to form real communities.

As we set off in my old pickup truck, I notice the almost complete absence of large industrial operations along the Columbia and Willamette rivers. The heavy industrial development has been replaced by a greenbelt. In some places, the riparian vegetation is in a natural, wild state and in other places it is manicured into parks. I make a mental note to find out what happened to the fuel storage tanks, railroads, and factories that once covered the river banks, but now I have to find my way out of Portland and head to Tillamook.

One hundred and fifty years ago the state of the salmon was grim and their future prospects bleak, so I am not optimistic about fishing prospects once we reach Tillamook Bay. After we leave Portland, I see the same recurring pattern of high-density neighborhoods surrounded by farms and what appear to be small industrial buildings. While still some distance from the bay, I see something that jumps out from the side of the road and throws me for a loop. A large sign says we are entering the Tillamook Salmon Refuge.

"Wow! A real salmon refuge; they actually did it!" I shout. Charlie gives me a sideways glance. "Charlie, do you know when they created this refuge?"

"Before I was born," he says. "Our class is going on a field trip to the refuge next year. We will spend a whole week learning about it."

We leave behind the clusters of high-density urban developments after entering the refuge. I see farms and evidence of timber harvesting, but they are very different from what I remember. The scale is smaller and the farms and logging operations appear to blend into the natural patterns of the landscape. Individual trees are selectively logged—no clear cuts are in evidence. Some forested areas appear natural while others are obvious tree farms. I am amazed by the size of the lush riparian areas along all the streams, even those flowing through the farms. It's obvious that humans and their economy are part of the refuge, but also that there are rules that soften the impact of human economic activities within the refuge.

I want to learn more about the refuge so we pull into the parking area at the headquarters building. At the door I see a uniformed biologist so we go over to him and I ask if he would mind answering a few questions. "That's what I am here for," he says. I have hundreds of questions and would like to simply say, "I'm from a different time so fill me in; tell me the whole story of salmon management since 2012." Instead I say, "I'm not from around here. When was the Tillamook Salmon Refuge created?"

"Well," he says, "the refuge concept has been around for over two hundred years and efforts to give refuge-like protection to the salmon and the rivers comprising the Tillamook watershed go all the way back to the late twentieth century. But those early efforts never gained serious political support. Finally, about 2020, some areas like refuges—they called them strongholds—were created, but those early strongholds bear little similarity to the salmon refuges we have today. Our current salmon refuges date from about 2060. That was the time of the great transformation when the country began moving away from fossil fuels as the primary source of energy. The first few decades of the transformation were difficult, but once the change took hold it created a cascade of secondary effects that changed all aspects of our economy, society, and culture. Salmon refuges were one of the side effects."

"How was the energy transformation related to salmon refuges?" I ask.

"It created a major shift in focus from a global economy to local or regional economies. The global economy collapsed without cheap oil to fuel its massive transportation system." He went on, "For example, when oil was cheap, we actually transported food thousands of miles from farms to markets. I remember reading that we imported food from countries like Chile, New Zealand, and China, and many others. Trucks and trains moved food and goods from coast to coast.

"The global economy contracted until it was replaced by something called bioregionalism or regional economies, which depended on the productivity of local ecosystems. Once the politicians realized that we had to depend on local ecosystems, it transformed attitudes towards resources like the Pacific salmon."

He pauses for a few seconds and appeared lost in his thoughts. "The transformation wasn't easy. There were some who just couldn't give up the old ways and wanted to go to war to get the last drop of oil, but we were fortunate that, by some miracle, at a critical point in our history, we elected a president and congress that had the foresight and courage to face reality. With the benefit of hindsight, it now seems obvious that we really had no choice and if we had chased after the last drops of oil it would have led to disaster."

I try to hide the fact that I am fascinated by the story. After all it is an old story to everyone but me. "But what about the salmon?" I ask. "How are they doing in their refuges?"

"Well, the runs are generally about 50 percent of their peak historical abundance—50 percent of their abundance before western Europeans entered the Pacific Northwest around 1850. However, the salmon are well above the impoverished levels during the years from 2000 to 2040 when they were at 5 percent or less of historical run sizes.

"The salmon were once one of the world's biological wonders. Yet for two hundred years after the arrival of Euro-Americans, our treatment of them was pretty shabby. It's a miracle any survived. What makes our management of salmon such a sad story is that the biologists had the information they needed to prevent depletion. Of course, we have much more detailed knowledge today, but the basic information to manage effectively was available as early as the 1930s. It was ignored. It is all there in the journals, reports and history books."

I say, "Yes, but there was too much political pressure to do the expedient, to stick with the status quo."

The biologist gives me a curious look and says, "Yes, there was political pressure. There is always political pressure. But in the past, managers, especially those in administrative positions, failed to separate politics from their responsibility to manage natural resources as a public trust. They used science when it furthered political, economic, and even personal agendas and ignored it when it challenged them. Today, we believe resources like the wild salmon are very precious. There is no excuse for what went on in the past."

"You seem to know a lot about the history of salmon management," I say.

He responds, "The history of fisheries management is a required course at the university, a full year. It's easier to avoid the mistakes of the past if you know what they were."

I could have talked to him all day about the history of salmon management between 2012 and 2150, but my time is limited and I have other burning questions. I ask him for more information about the refuge.

"It's organized like an onion in concentric layers. The core areas are the critical habitats and stream reaches. Those areas are protected and left undisturbed by human activities. Reaching out from those core habitats are the onion-like layers, each with a different level of protection, and different kinds and scales of human activities. The last layer of the refuge reaches out to the edge of the catchments of the five rivers—the Miami, Kilchis, Wilson, Trask, and Tillamook—that flow into the bay. The Tillamook Salmon Refuge recognizes that the ecosystems of the Pacific Northwest are really natural-cultural systems, which means that human activities and economies must be part of the refuge. However, within the refuge boundaries those economic activities are regulated to achieve the primary purpose of maintaining an ecosystem that supports a productive, salmon-sustaining ecosystem. All the salmon refuges—the Rogue, which includes the streams from the Elk and Sixes rivers to the California border, the North Umpqua, the Salmon, the Lower Columbia, the Olympic, and the Skagit—all use the same approach."

Out of the corner of my eye I notice Charlie is getting restless so I ask the big question. "Do you allow fishing in the refuge?"

"Yes, the salmon refuges were never intended to be museums. The refuge program would have failed if the salmon runs remained too small to harvest. The fishery has four elements: a small commercial fishery, a sport fishery for locals—those who live within the refuge boundary— and a sport fishery for people who live outside the refuge. Fishing permits for the outsiders are acquired through a lottery held twice a year. Finally there is a daily drawing for a small number of permits. If you are interested in fishing, the daily lottery is your only hope. Sign up at the desk. The drawing for tomorrow's fishing is at 5 p.m. and it is limited to five permits." The biologist looks at Charlie and says, "You will both have to register for the drawing."

After we register, I ask Charlie, "How about getting something to eat, then we can look around the refuge before the drawing?" We stop at the first café we come to in Garibaldi. The menu is very interesting. It is the fall menu, and the list of dishes reflects the seasonal availability of various foods. Each item on the menu names the farm or fisherman that supplied the main ingredients, and the suppliers are all local. We order fried oysters with a big glass of milk. I tell the waitress, "Your menu listing the source of food is amazing. I like it."

She says, "You aren't from around here, are you? Nearly all the restaurants in the Northwest give that information."

After lunch, we drive around the refuge, stopping at a place where we watch several fall Chinook holding in the river. I tell Charlie that the salmon will soon be spawning and laying their eggs in gravel nests called redds. We stay there for a couple of hours quietly watching; the whole time Charlie is fully absorbed in watching the fish. I am amazed at the kind of questions he asks. Is that bird a water ouzel? He names the nearby trees and asks if his identification is correct. It is obvious he knows a lot about the stream and its flora and fauna. During our vigil along the river bank, my emotions move between wonder at what has been accomplished, satisfaction that finally people woke up to the importance of place, and a deep sense of disappointment that in my time we were so shortsighted.

At 4:30 we return to the refuge headquarters. The biologist I talked to earlier comes over and says, "I see you are back for the daily drawing."

I answer, "Yes, but I didn't expect this crowd. Are all of these folks here for the drawing too? There must be a hundred or more."

"People in the Northwest love their salmon and salmon fishing," he says.

"What about hatcheries? Do you have hatcheries within the refuge?"

"No, the refuges were established to nurture natural salmon production systems. The attachment to hatcheries killed the earlier attempts to create salmon refuges. People back then were reluctant to create salmon refuges because they believed that artificial propagation could replace habitat. It was about 2040 that they finally faced the reality that, in many ways, hatcheries were part of the problem."

"So were all the hatcheries closed?"

"No, a set of rigorous criteria and standards for hatchery operations were established, and the hatcheries that couldn't measure up were closed—and that was most of them, but there are a few still in operation. Today, they are no longer used in a way that fisheries historians call symbolic stewardship; that is, they are no longer used to give the appearance of stewardship while in reality their purpose was to avoid it."

"Well, it looks like they are about ready for the drawing. Good luck."

Neither Charlie nor I draw one of the five winning numbers. After another look at the displays in the headquarters we start back towards Portland. I tell Charlie I am surprised at how much he knows about the plants and animals along the stream. I am even more surprised by Charlie's response. My statement triggers a cascade of information about the salmon, rivers, plants, and animals. He thoroughly understands the salmon's life history and habitat requirements and even the state of the streams and salmon near his neighborhood.

"Where did you learn all this?" I ask.

"From my teacher," he says. "Mr. Edwards, my teacher, said that we are all part of a community, the birds, trees, every living animal and plant and especially the salmon. To be a good citizen in the community, we have to not only know the names of the other members of the community, but where they live and what they need to survive. We have to be considerate of the members of the community and help them when they need it just like we help other people in our neighborhood. We all live here; it's our home."

Chapter 6—Salmon, People, and Place

"Goodbye," said the fox. "And now here is my secret, a very simple secret: It is only with the heart that one can see rightly; what is essential is invisible to the eye."

—Antoine De Saint-Exupery[1]

[No] big things happen for just one reason. This may sound like a trivial matter, but in fact it's quite important. Notwithstanding the attraction of tidy, sound-bite-ready, just-so stories, most things happen not for one reason, but for many. … The more one looks at any situation, the more one marvels at the interlocking gyre of cause and effects. Part of the beauty of the world, even including its disasters, is its complexity.

—William deBuys[2]

Of Rivers and Salmon

I am a salmon biologist and I consider myself fortunate to live within sight of the Columbia River. It was once one of the world's great salmon rivers. I often turn away from the computer, look at the river, and think about its problems and the problems of all salmon rivers. I am thinking about the river today; it is March 15, 2012. The spring Chinook are returning to the Columbia, keeping alive a ritual that has persisted for many thousands, if not millions, of years. The spring Chinook swimming up the river migrate hundreds of miles from oceanic feeding grounds to reaffirm their attachment to a special place, to the stream reach where their life's journey began. I can't see the salmon, but I am confident they are there, in the muddy waters of a spring freshet. While I look out at the river, I wonder how long the people of the Pacific Northwest will look at their rivers in spring and be confident that wild salmon are on their way home.

The spring Chinook will disperse to their home tributaries in those parts of the basin not blocked by dams and, in six or seven months, they will spawn. After spawning, they have one last task to perform to fulfill their evolutionary mandate to preserve their species. Once the incubating eggs are safely buried in the stream's gravel, the adult salmon die. By their death, they give their bodies to feed and enrich the entire ecosystem.

Over millions of years the river and the fish became enmeshed in a web of relationships that developed at the slow pace of evolution, giving the whole salmon-sustaining ecosystem the resilience needed to cope with change. Each generation of salmon returning to the rivers of the Northwest confronted change; gradual evolutionary change, and change due to large events such as volcanoes, landslides, and fires. Those latter changes gradually softened and then blended into the web of relationships.

Salmon returning to the rivers today are confronted with changes of a very different nature. The changes occur in rapid succession and they are big and permanent. Massive walls of concrete block the path of the salmon's migration. Dewatered rivers, high temperatures, silt, and polluting chemicals are no longer episodic events, but permanent conditions. The rate and magnitude of change is outstripping the salmon's capacity to coevolve complementary adjustments to the new conditions. The web of relationships, which acted as a safety net against extinction, is frayed with gaping holes, leaving wild salmon dependent for their survival on the flimsy protection of the federal ESA.

Eventually, but with lightning speed when compared to evolutionary change, we replaced rivers with things that still look like rivers to the humans who engineered them, but to the salmon, are impoverished substitutes. The destruction of the once boundless capacity of rivers to produce salmon was justified, made rational, and accepted as inevitable by the myth that says it's possible and even desirable to replace wild salmon with an artificially propagated substitute. The nurturing ecological relationships—the things we can't see, but are so critical to the salmon's survival and the health of the ecosystem—are being degraded and stressed to the breaking point. But we are told, "Do not worry, because fish factories are a more than adequate substitute." The few remaining wild fish are becoming aliens in their home streams. They hang on to what little habitat remains like the thin line of little oxalis plants in the clear-cut forest.

This is not the inevitable or unavoidable consequence of reality or any of the other words used to convince us to accept the status quo. We can choose to enter into a different relationship with the salmon and their rivers. We can set ourselves on a different path to the future, a path that veers away from the impoverished legacy that future generations are sure to inherit, if we continue to accept the status quo. As noted previously, fisheries administrators and political leaders derive too much comfort from the status quo to expect them to lead the search for an alternative. It will take significant pressure from the public. The previous chapter laid out a list of changes in salmon management that, if implemented, would begin the move toward a new salmon story and a new approach to their management. But real change will take more than a to-do list. Change must be embedded in a different way of thinking about the Pacific Northwest and what it means to call this place home.

A Shared Vision of the Future

The answers to these challenges will be plural, not singular, but no answer will count for much—will, in fact, be an answer—if it is not backed by strong social will and collective commitment.

—William deBuys[3]

I know it sounds presumptuous to suggest that the fate of a fish should trigger the development of a shared vision of the future for the Pacific Northwest. I believe there are reasons why the salmon hold a prominent role in this endeavor. First, it's hard to imagine a shared vision for the Pacific Northwest that ignores this regional icon. Wild salmon penetrate the ecosystems of the Pacific Northwest so thoroughly that they come into direct or indirect contact with nearly all human activities. As a result, a shared vision based on the persistence of these magnificent animals must consider the many different values and aspirations held by Northwesterners. Because the salmon's annual return to the rivers is a place-defining event and because the salmon penetrate the landscape so thoroughly, a shared vision for their recovery will approximate a shared vision for the entire bioregion.

Second, a shared vision must include consideration of the commons. The commons include, but are not limited to, natural resources such as wild fish and wildlife, publicly owned forests, rivers, wilderness areas, parks, the air we breathe, and ecosystems. It can also include infrastructure built by humans, such as roads, publicly owned water and sewage systems, fire and police stations, and ports. The commons are those constituents of a place that are owned by all of us so they must play an important role in a shared vision for the future.

In a culture that places a high priority on the unfettered pursuit of individual and corporate wealth, caring for the commons receives little attention. In fact there are many people today who would sell off the commons to private interests. The pursuit of individual gain at the expense of the commons gave rise to the fragmented management of ecosystems described earlier in the book. Having lost sight of the commons, we do not fulfill our collective responsibility for its care. The current state of the salmon is a consequence of that neglect.

Achieving a shared vision won't be easy. The bureaucratic boundaries that fragment ecosystem management and serve the narrow purposes of special interests will be a major impediment to the development of a shared vision. It is a major but not insurmountable problem. The fragmented management of ecosystems and all that it implies is a human construct and humans can change it. But to make that change requires leaders who recognize the need and have the resolve to challenge the status quo. It will take pressure from an informed public to help our political leaders see the wisdom of dissolving the bureaucratic boundaries that fragment the management of the Pacific Northwest's ecosystems.

The listing of Pacific salmon under the federal ESA sent a strong message: the regional icon is on the brink of extinction. An aroused public wanted something done. Some wanted to save the icon. Others wanted to avoid the economic penalties that were sure to follow the listing. The massive funding triggered by the listings—funding that has been forthcoming for at least twenty years—is evidence of the strength of the public commitment to salmon recovery. Because the motivation to recover the salmon springs from widely different values, it has the shape and feel but not yet the coherence of a shared goal. But concern over the fate of the salmon could be used by our political leaders as the beginning of a public conversation about a shared vision.

Most political leaders in the region have shied away from an honest conversation about the salmon's problem. Oregon Governor John Kitzhaber stands out as the exception. During his first term, he called for a public conversation about the fate of Pacific salmon, a conversation that recognizes that successful recovery of wild salmon requires broad public participation. According to Kitzhaber: "Eventually it will take sacrifice—individually and regionally. But if shared, the load will be bearable. The choice before us is clear: either we as a society choose to make sacrifices—or choose to sacrifice salmon. I do not believe we will choose the latter path."[4]

Governor Kitzhaber has repeatedly called on the region's leaders to begin the difficult task of starting a public discussion leading to a shared vision of the future.[5] He is not naïve in this proposal; and he realizes the difficulty. In a speech to Trout Unlimited he warned: "What I propose will not be easy—and I speak from experience. I have been swimming up this stream for the last six years." In a magazine article in 2006, Kitzhaber called for nothing less than a paradigm shift (a new story) that replaces the perpetual confrontation with a "place-based consensus process." We have all seen political leaders and agency administrators perform the obligatory hand wringing over the fate of the salmon. Their concern, however, is not so strong that they have accepted Kitzhaber's challenge to begin a public conversation about a shared vision and shared sacrifice.

Governor Kitzhaber recognizes that salmon recovery needs more guidance than the oft-repeated and inane "double the run" goals and recovery plans constrained by the self-imposed boundaries of fragmented ecosystems. It will take a change in the status quo, change that is shared across the bioregion. How much change is required to ensure the persistence of wild salmon? I can't answer that: it is tied to the region's shared vision for the future and the role of salmon in that vision. If the shared vision sees the future of salmon only as a commodity produced in fish factories, then very little change is needed. On the other hand, if the shared vision includes a future with healthy rivers where wild salmon populations are sustained at fishable levels (but not a return to pristine levels of abundance), then substantial changes in our relationship to rivers and their salmon are needed. How far the region will move from the current state of its rivers and wild salmon is a decision that should be made after a public weighing of tradeoffs among all those who call the Pacific Northwest home.

What kind of Pacific Northwest do you want for your children and grandchildren? Do you want to play poker with the fate of the place where your children and grandchildren will live and just accept the hand you are dealt, especially if the dealer is manipulating the cards in favor of special interests? Initiating a public conversation about a shared vision will take pressure from the public, a lot of pressure. It will take support for those leaders willing to take the politically dangerous step to initiate a real public conversation. At the same time it will require public criticism for those who shirk their responsibility. Speak out, get involved in whatever way suits your circumstances. Let your leaders know the status quo must be changed.

To arrive at a shared vision of the future, we will have to consider several different ways that people value ecosystems and the natural resources they produce. Many different points of view will be considered. I suggest that among the points of view considered should be one borrowed from ancient Native American culture. I am referring to the belief that natural resources like the salmon are gifts.

The Salmon's Gift

Is each salmon only a chunk of cash to be divided up by bureaucrats?
—Tim Bowling[6]

Whatever your belief about creation and the origins of life on Earth, that belief, unless it is completely overridden by the single-minded pursuit of profit, should include respect for nature's gifts and a sense of wonder at the diversity of life that blankets our planet. Whatever drives your relationship with the natural world— whether you cut trees or build houses out of the wood, irrigate a farm or cook the food it produces, whether you push barges up a river or catch salmon, whether you convert sawdust to paper or use paper to write books—you are surrounded and sustained by nature's gifts. The annual return of the salmon to the rivers of the Pacific Northwest is one of those gifts. A gift in this context is not the same as gifts received at Christmas or on a birthday. When I say the return of the salmon is a gift, I am using the ancient meaning of the word. Here is how I described this meaning of the gift in *Salmon Without Rivers*.

> *In ancient economies in such disparate parts of the world as Southeast Alaska, Polynesia, and early Germanic Europe, the gift formed the basis for exchange and commerce. … The very act of accepting a gift meant the receiver also accepted an obligation to return it in kind to the giver. Rather than transfer goods through a market, the gift-based economies*

worked through a cycle of obligatory returns. The giving of gifts generated a record of obligations that defined the credit structure of society. In ancient economies, the gift and the obligation it incurred were more like a bank loan and its repayment than like our exchange of gifts at Christmastime. Anthropologist Marcel Mauss equated the gift in ancient economies to Adam Smith's "invisible hand" in modern market-driven economies.

We enthusiastically accepted the gift of salmon, but failed to treat it with the respect it deserves. We failed to meet our obligation to return the gift in the way that only humans can. We failed to return the gift of salmon with the gift of stewardship.

One hundred sixty years ago, Euro-Americans, using their Western culture, economy, social structure, and science displaced Native Americans in the exploitation and management of Pacific salmon. It then took more than a hundred years for Native Americans to gain the status of co-managers of the salmon. Euro-Americans made many changes in the relationship between humans and natural resources in the Pacific Northwest, but perhaps the most fateful was the change in overarching metaphors from gifts to money. That switch in metaphors converted salmon to a commodity, which when combined with a mixture of greed and hubris, depleted or drove to extinction many wild salmon populations south of British Columbia.

Euro-Americans inadvertently proved the truth of the warning embedded in the Native American myth of the five houses of salmon: if humans do not treat the Salmon King's gift with respect, he will withhold that gift. The narrative of that myth is a fiction, but it's hard to deny the reality of the lesson. Instead of returning the gift of the salmon with the gifts of respect and stewardship, we have polluted, dewatered, and dammed rivers, covered the spawning gravel with silt, denuded stream banks of riparian vegetation, elevated stream temperatures, overharvested the salmon, and, to top it all off, treated salmon as widgets produced in factories. Consistent with the Euro-American's metaphor for resources, those abuses were justified because they made money.

When money is the overarching metaphor for natural resources, it leads inevitably to economism as the sole means of assigning value to those resources. Economism is the belief that: "Economic evaluation is the single comprehensive method for determining all values including environmental values and, once commoditized and given monetary value, the exploitation of natural resources and the tradeoffs between the environment and development are treated simply as market transactions."[7] A clear sign that the malady of economism has infected salmon management programs is the use of economic metrics as a surrogate for the ecological health of wild salmon in the Oregon Department of Fish and Wildlife's budget request (see Chapter 4).

The underlying assumption of economism is that every part of nature has a price; we should trust economists to set the dollar value and then trust the future of resources like salmon to the magical, invisible hand of the market. Author Bryan Norton argues that we should avoid the use of a single, comprehensive measure of environmental values and accept a pluralistic approach.[8] Norton's recommendation echoes Governor Kitzhaber's call for a shared vision, a vision incorporating the different ways citizens of the Northwest value salmon. A poll commissioned by *The Oregonian* newspaper in 1997 revealed pluralistic public opinions on the value of salmon. When Oregonians were asked to give their reasons for preserving salmon they responded this way:

36%—Because salmon are part of Northwest history and heritage
35%—As a gage of water quality and the environment's health
9%—For sport fishing
8%—Just to know they are there for personal or aesthetic reasons
6%—For commercial fishing
2%—I don't care about preserving salmon runs
4%—Don't know or no response[9]

Only 15 percent of those polled said that they wanted to preserve the salmon for the commercial or sport fisheries, reasons that lend themselves to economic values, while a total of 79 percent cited reasons that are not easily converted to a monetary value. Dollar bills and markets alone—economism—are not adequate measures of the diverse ways the public values wild salmon. The people polled did not view wild salmon as simply chunks of cash for managers to divide between the sport and commercial fisheries.

Another shortcoming of economism as the sole arbitrator of environmental and resource value is that it distorts our sense of place. It "diminishes the stability of a place-based community and attachment to local resources and place-based practices."[10] When salmon become commodities produced in factories, they do not demand that we return their gift with the gift of stewardship, they only demand that the right amount of money be laid on the counter.

Economic factors are an important component of fishery management, but when they dominate decision making, the impoverishment of the fish, the fishermen, and dependent communities is guaranteed. Economism has contributed to the depletion of fish stocks all over the world.[11] I believe that thinking of wild salmon as a gift will reconnect people and salmon management to place. It will cause us to pay attention to how we are altering place and show us the error in producing ecologically placeless salmon in an industrial production system. Finally, it will remind us of our obligation to return the gift in the only way we can—with informed stewardship.

Can a society that lays claim to modernity adopt the gift as a metaphor for natural resources such as salmon? The author David Griffith, in writing about estuaries on the mid-Atlantic coast, says it is not only possible, but highly desirable. Griffith believes that our treatment of natural resources and the impoverishment of the local resource-dependent communities are two sides of the same coin. The development, commoditization, and sale of coastal ecosystems based solely on the rules of economism degrade ecosystems. The resort condominiums, boutiques, and marinas that service yachts devour habitat and natural productivity. Sport and commercial fishermen and the communities they support cannot compete with condominiums and yachts when the dollar is the only measure of value. When we give in to economism's single-minded quest to give a dollar value to everything, then sell it in the market to the highest bidder, we toss aside the fish, fishermen, and their communities as the inevitable consequence of progress. Griffith argues that the way we treat each other and the way we treat nature should be measured against the standards and obligations that govern the giving and receiving of gifts.[12]

I am sure that some who read this discussion of the gift will say it's a naive approach to a serious problem. Those who are firmly anchored to economism and its industrial production system will be especially critical. After all, treating salmon as gift violates all the assumptions of command-and-control management. It has none of the security of a firm dollar value with which to weigh management decisions. Converting commodities to gifts attaches too many obligations and too many ecological complexities to salmon management. The gift would obligate us to stewardship, which is what we have claimed to be doing for over a century.

But thinking of salmon as a gift doesn't mean we have to abandon their value as a commodity. It simply means that we will have to get serious about accepting our responsibility as stewards.

A Coevolutionary Perspective

Like the wider concern for the environment, the protection of ecosystems endangered by what used to be called "progress" reminds us that coevolution may be the way of the future as well as the past. The tempo and scale of life on the planet is biological at its base. To the extent our technologies and organizations can heed and harness those rhythms, our species may prosper. But we ignore them at our peril.

—Kai Lee[13]

Since coming into existence four billion years ago, the fragile thread of life on earth has replicated itself and diverged into wondrous new forms whose life histories are continuously adapting to the attributes of place. Webs of relationships among the many forms of life give strength and resilience to this living tapestry. At various points in the landscape, those relationships coalesce into nodes we call ecosystems. The species and their life histories we see today are not the end of life's four-billion-year journey, but the latest chapter in a very long story. The trajectory of that story is guided by a species' inherent genetic possibilities, and by the strength and nature of its interactions with other species. Bound together by their relationships, species don't just evolve—they coevolve.

How important is the concept of coevolution to the recovery of Pacific salmon? Two or more species are said to coevolve when their relationship is so strong that an evolutionary change in one results in a complementary change in the other. Some coevolved relationships are highly specialized and intense. For example, the white Madagascar orchid and the hawkmoth coevolved extremely specialized adaptations. The hawkmoth has a nine-inch proboscis, and it uses this unusually long appendage to feed on nectar at the base of a tube of similar length hanging from the orchid.[14] The orchid gives the moth the gift of nectar, and the moth returns the gift by pollinating the orchid. In this extreme case, Gary Nabhan's statement about the importance of relationships comes to mind. If the relationship between the hawkmoth and orchid were lost—if the moth were extirpated—the orchid's future would also be in jeopardy. Not all coevolved relationships are so specialized or intense; however, they are so important that biologist John Thompson believes that we cannot understand the diversity of species unless we understand the diversity of the coevolved interactions among them.[15]

For several thousand years wild salmon were a keystone species in the ecosystems of the Pacific Northwest. They occupied a special place at the nexus of relationships that tied together ocean and freshwater, terrestrial and aquatic ecosystems. When the salmon returned to their home streams they carried with them, in addition to the next generation of salmon, nutrients acquired on the oceanic feeding grounds. Those nutrients, released from the decaying salmon carcasses, nourished terrestrial and aquatic food webs from the headwaters to the estuary.[16] When the wild salmon ran upstream, they transformed a river into a great table, set with a feast that nourished the entire ecosystem: cedar trees, bears, eagles, humans, and many others.[17] For more than ten thousand years the species drawn to this feast were connected to the wild salmon through a web of coevolved relationships and, as Gary Nabhan pointed out, those relationships were an ecological safety net that kept species from falling into the abyss of extinction. Today, because habitats are routinely degraded and wild salmon runs are depleted, that safety net is full of holes and the salmon, among other species, are on the verge of slipping through.

Among the relationships that define the salmon's evolutionary trajectory, the human-salmon relationship stands out as the most important because humans have the power to willfully alter the course of that trajectory. It is a power that is often wielded with a hubris and an incredible level of ignorance about what it is we are changing and the consequences of our actions.

Our stories determine whether we will coevolve a sustainable relationship with natural resources such as salmon or whether we will destroy relationships and extirpate those resources. In that regard, it is important to remember that we don't just tell stories—stories about rivers and harnessing their power and stories about salmon factories. We tell these stories; but these stories also tell us.[18] Stories dig deep and reveal the beliefs, assumptions, and myths that determine how we comprehend and interact with the world around us. If we pay attention to and approach our stories with an open mind, they will tell us important things about ourselves, things that are buried so deep inside us that they remain hidden like the hidden mass of John Livingston's iceberg. The old adage—out of sight, out of mind—applies here.

What our salmon story says about us can be summarized in two words: command and control. We commanded that the threads related to wild salmon woven into the four-billion-year-old tapestry of life be pulled out and engineered so they could be controlled to meet economic expectations.[19] Our story of resource management is a story of command and control. In an important paper, biologists "Buzz" Holling and Gary Meffe described the structure and consequences of the command-and-control methodology in natural resource management.

> *The command-and-control approach implicitly assumes that the problem is well-bounded, clearly defined, relatively simple and generally linear with respect to cause and effect. But when these same methods of control are applied to a complex, nonlinear and poorly understood natural world, and when the same predictable outcomes are expected but rarely obtained, several ecological, social and economic repercussions result. ... Our thesis is that adoption of such command and control has resulted in a pathology that permeates much of natural resource management and precludes long-term sustainability.*[20]

Implementing command-and-control resource management often yields initial, short-term successes, which may be real or, in some cases, wishful thinking. For example, when success eluded early fish factories, it was created out of whole cloth, at first by simply claiming that the large, wild runs in the early decades of the twentieth century were made up of factory-produced fish. This was a safe assertion because at the time it was impossible to check its veracity. Later, as the wild salmon runs were depleted, it became obvious that the fish factories were failing to maintain or increase the supply of salmon. To maintain the illusion of success, the measure of hatchery performance shifted from the number of adult salmon supplied to the fishery to the number of eggs taken in the hatchery or the number of juveniles released into the river.[21] Short-term success achieved through command and control, whether it is real or imagined, infects the management institution with a pathology. Here is how Holling and Meffe describe it:

> *Priorities thus shift from research and monitoring (Why 'waste' money studying and monitoring apparent success?) to internal agency goals of cost*

efficiency and institutional survival. The second feature of the pathology thus emerges: growing isolation of agency personnel from the systems being managed and insensitivity to public signals of concern—in short, growing institutional myopia and rigidity.[22]

Institutional myopia and rigidity resonates with my experience. For example, before the paper "Salmon at a Crossroads" was published, Willa Nehlsen sent a review draft to several biologists in the Pacific Northwest. I described the reaction of one biologist in Chapter 2. He definitely exhibited signs of myopia. But in addition to his response, I received letters from other managers criticizing the paper, not because it contained technical errors, but because releasing the information on the miserable state of the Pacific salmon would, as one letter writer put it, allow some to create "mischief." Some saw "mischief" in describing the state of the salmon because it was powerful evidence of the failure of the status quo. For the record I received many, many positive comments from biologists. The positive comments largely came from field biologists, those in intimate contact with the rivers, salmon, and ecosystems. The critics were largely office-bound administrators.

For more than a century, salmon management has been based on a command-and-control regimen. The resulting pathology reduced life history and genetic diversity and created an institutional myopia and rigidity that either ignored failure or made excuses for it. Instead of trying to design a management approach that was in harmony with the "tempo and scale"[23] of the seasonal patterns in the ecosystem and the salmon's life histories, managers created a model of a highly simplified industrial production system. They then created an engineered facsimile of the salmon to fit the industrial model. Adhering to this model and to its command-and-control approach made a coevolving relationship with wild salmon impossible.

In Kai Lee's essay on the salmon's problem, he tells us that coevolution, which was part of the human-salmon relationship in the past, may also be the future of that relationship. That is, if we deem wild salmon important enough to have a future. Managing with a coevolutionary perspective is essentially a form of adaptive management. It is the recognition that salmon management is one long experiment and that managers and management institutions must continuously learn from the results of the experiment. In a coevolutionary approach, history matters, and managers must have a working knowledge of the history of the human-salmon relationship. Superimposing a simple, industrial-production system onto real ecosystems terminates rather than nurtures a coevolved relationship with wild salmon. A coevolving relationship respects the integrity of place.

☙ ❧

I devoted this chapter to a shared vision that incorporates multiple points of view and values, to the Native American idea that salmon are a gift, and to the need to return to a coevolving relationship with wild salmon and other natural resources. Call these ideas the beginnings of an ethics of place. It is an antidote to the complacent acceptance of the myth that it is acceptable to substitute an industrial salmon production system for healthy ecosystems and wild salmon. A shared vision, treating resources as gifts and striving for a coevolved relationship with this place and all its native species, if given serious consideration, will put us on the path toward a new salmon story—a salmon story that respects and nurtures the ancient connections between salmon, people, and place. It will set us on the path toward what some have called Salmon Nation.

Salmon Nation: A New Vision for the Pacific Northwest Bioregion

*And most important, only when governments that typically ensure economic interests and values over all others decide that **they** [emphasis in the original] are willing to re-construct the human-salmon relationship as an ecological one rather than an economic one will the true salmon wars, the wars between society and the salmon, ever be over.*

—Rik Scarce[24]

Salmon Nation: Two words that many of us who live in the Pacific Northwest have seen on book covers, bumper stickers, T-shirts, posters, billboards, and on a banner in at least one restaurant. But Salmon Nation is more than an advertising or sales gimmick. If the annual runs of salmon were the place-defining event that I claimed in Chapter 1, then Salmon Nation are the words that could remind us of the event and could reaffirm its importance to a healthy sense of place. They could, but they haven't—yet. Salmon Nation has the potential to correct the distorting effects of economism and allow us to see some important things in a different light, things whose value comes not from our wallets but from our hearts.

Salmon Nation reminds us that wild salmon are an important resident in the Pacific Northwest bioregion, reminds us of their long history here, and their ecological significance as a keystone species. Salmon Nation is also a perspective. It's a bioregional perspective that says we in the Pacific Northwest recognize and hold dear our region's unique attributes and gives one of those attributes, the wild Pacific salmon, the status of icon or totem animal. Wild salmon are both symbol and spirit of the region. Citizens of Salmon Nation recognize that wild salmon are a gift from the region's ecosystems and when we accept that gift we take on an obligation to return it with stewardship.

Salmon Nation defines a bioregion whose boundary extends, in the words of Timothy Egan, to "any place the salmon can get to."[25] It's a boundary that tells the world that we will accept what it has to offer, if it doesn't put at risk what we value. It's a permeable boundary that needs to be monitored to ensure that what enters or leaves is compatible with the region's shared values and vision. It's a boundary that recognizes the global economy, but on terms that protect what we consider unique and precious. Salmon Nation does not accept the need for global homogenization to achieve the myopic goal of economic efficiency. Salmon Nation recognizes that, as the era of cheap fossil fuel winds down and the global economy contracts, local cultures will of necessity return to a coevolving relationship with their ecosystems and depend more and more on the productivity and gifts of the bioregion. Salmon Nation says that the citizens of the Pacific Northwest recognize the importance of retaining the ability to craft solutions to problems based on the natural uniqueness, attributes, and resources of this place. There is a lot packed into the two words Salmon Nation.

If we allow Salmon Nation to break free of its current status as a sales gimmick it could evolve into what George Lakoff calls a deep frame;[26] a way of thinking and seeing that significantly changes behavior toward the natural attributes of the bioregion. Today, Salmon Nation is a long way from a deep frame, it hasn't become part of our conceptual infrastructure, but, even though it hasn't escaped the status of a slogan, it has the potential for much more. Look inside Salmon Nation and you will find a shared vision of the future, a rich array of gifts from ecosystems and a web of coevolving relationships. These have been there all along, but for the past one hundred and fifty years they have remained invisible, below our cultural radar. It's time to return to a coevolving relationship and reclaim our place as stewards and citizens of Salmon Nation.

Aldo Leopold's Conceptual Foundation Revisited

Rational salmon management is not just a search for technologies: it's a search for values.

—Dennis Scarnecchia[27]

More than sixty years ago Aldo Leopold's book *A Sand County Almanac* was published. Through this book and his many essays, Leopold became one of the most influential spokespersons for conservation. Toward the end of his life, he penned the essay "The Land Ethic," which is an important part of Leopold's conceptual foundation. Most of us who care about nature and who work toward conservation's ideal—a harmony between humans and nature— view "The Land Ethic" as a summary of conservation's core beliefs. We have read and been inspired by Leopold's core ideas, but have been slow to assimilate those ideas into our intellectual frames. I am afraid Leopold would see little of his view of conservation being put into practice in fish and wildlife agencies today. The dominance of economism and economic measures of performance were rejected by Leopold long before Bryan Norton revived that idea. While Leopold still receives our praise for his contribution to conservation, we have all but forgotten the basic principles of his land ethic.[28]

In "The Land Ethic," Leopold describes a cleavage of views among biologists and specialists working for conservation organizations such as fish and wildlife agencies. One group views ecosystems as organic machines producing commodities, and they use economism to measure the value of that production. The other group assigns a broader function to the ecosystems. Leopold was vague on just what that function was. Remember he wrote more than sixty years ago. Looking at the cleavage today, I believe, with the benefit of hindsight, that the second group views the ecosystem as a community of beings exchanging gifts through webs of coevolving relationships. The cleavage Leopold describes has been evident in fishery management since the early decades of the last century.

Leopold also tells us that all ethics are based on the idea of community and the need for cooperation among members of the community. To achieve an ethical basis for our relationship with nature, the "land ethic simply enlarges the boundaries of the community to include soils, waters, plants, and animals or collectively the land."[29] We must adopt Leopold's expanded definition of community to accept the idea that natural resources are gifts. Accepting the gift of salmon implies a relationship based on a set of obligations—in other words a relationship based on a land ethic. The key to adopting a land ethic must, according to Leopold, start by abandoning the dominance of economism.

Aldo Leopold and his ideas illustrate a paradox. We recognize the validity of those ideas, especially the land ethic, and as a consequence, we honor the man and acknowledge his unique contribution to conservation. On the other hand we have been too timid to actually put those ideas into practice, to embed them in the stories that guide the decisions of resource managers. We have not used Leopold's lessons to devise standards by which resource managers are held accountable. Salmon restoration based on the old story, the story that has contributed so much to the creation of the salmon's problem, will continue to fail to achieve recovery, just as it has failed for over a century to prevent decline. To put our relationship with the salmon on a positive path, we do not need more restoration derived from the old story. We need re-story-ation. We can't expect to get a different result in our restoration activities until there is real change in the story, in the conceptual framework.

There are those who will say that regardless of the need for change, especially the need for cultural change, the task is too big, too difficult. I disagree. Instead of keeping the ideas of pioneers like Aldo Leopold and Rachael Carson locked away in an intellectual closet, to be taken out on special occasions such as Earth Day, we need to keep them in the conversation and in the debates. Over time, those ideas will enter into our deep frames and change our story. Twenty-five years ago recycling was a kooky idea that a few people on the fringe of the conservation movement practiced and preached. By keeping recycling in the conversation, recycling enthusiasts changed the way we think about trash and precipitated a cultural shift of significant magnitude. It has become a regular practice for many citizens to recycle, reduce, and reuse, with a well-recognized recycling logo. It works.[30]

In developing a new story, we need to recognize that the real experts on salmon recovery are the salmon. They have the accumulated lessons from millions of years of evolutionary experience. The Pacific Northwest is one of the most geologically active regions of North America. That means the salmon coevolved in a landscape subjected to mountain building, volcanic eruptions, landslides, gigantic lava flows, earthquakes, and fires, all of which could destroy or block access to habitat. The salmon survived through the millennia by adapting life history strategies that fit the evolving habitat. But their long evolutionary history did not prepare them for the Euro-American worldview or for command-and-control management and the aggressive attempts to mold salmon-sustaining ecosystems to fit the Euro-American vision of a simplified-industrial model. So far, in the relationship between Euro-Americans and the salmon, the salmon have been expected to change their evolutionary legacy to fit the newcomers' story of factory-produced salmon. We have acted as though the salmon are the newcomers who must adapt. That many citizens of the Pacific Northwest, including many salmon advocates, do not see that this is a backward approach is testimony to the power of stories.

If Thomas Berry is correct in his assertion that our biggest crises occur when our stories no longer protect the things we value, then salmon biologists and managers are working in a continuous state of crisis. The institutions they work for are locked in a story and a behavioral context that have failed to protect the salmon, have created a crisis, and are now prolonging it. Salmon managers seek some relief from the crisis by shifting the baseline forward. Natural variation periodically produces a bump in the abundance of salmon. Those bumps and a shifting baseline allow managers to spin a positive narrative. The year 2012 was one of those occasions. After a series of years with favorable ocean conditions, the fall Chinook run into the Columbia River has responded with an increase in abundance.

With a predicted run size of 654,000 fall Chinook, salmon managers on the Columbia could boast that the run would be 106 percent of the previous year's run and 113 percent of the average run sizes for 2002 to 2011.[31] Notice the shifted baseline. The predicted run size is actually 23 to 39 percent of the range of historical run sizes developed by the Northwest Power Planning Council.[32] Have we reached the point where salmon management relies on the vagaries of climate and the denial of history to manufacture positive news?

Several years ago, when I was the assistant chief of fisheries at the Oregon Department of Fish and Wildlife, I taped a three by five card to the wall next to my desk with this written on it: *If the salmon runs next year increase it is the result of our programs, but if they decrease it will be due to changing ocean conditions and beyond our control.* The statement was a tongue-in-cheek parody of real comments I had heard in halls and conference rooms of management agencies or read in the newspapers. These words are an example of the convoluted mental gymnastics needed to build a positive narrative for a story that is no longer in touch with reality and no longer protects what we value, the wild Pacific salmon. My goal in writing this book was to loosen the hold of the myths, beliefs, and assumptions about nature that bind us to the failed story. It is a task that is too difficult for those who work within the management institutions to take on by themselves. It will take prodding, questioning, and help from a public that understands the nature of the problem. And there are organizations working today to provide that information. The Wild Fish Conservancy and the Native Fish Society are two organizations that have been fearlessly informing the public about the importance of changing the status quo.

The information those organizations so willingly share will resonate with and be useful to those folks who are prepared to receive it. Getting prepared is as simple as paying attention. If people pay attention to the place where they live, they will see evidence every day of the tenacity of this four-billion-year-old gift we call life. The way life in all its varied forms stubbornly survives, even under the harshest and most thoughtless treatment by humans. Over my career I've seen too many examples like the oxalis plants on that logged-off hillside to agree with those who say stop trying and accept the inevitable, or the inane statements of those who believe that life in its wondrous forms can only be nurtured by the invisible hand of the market, or that Pacific salmon need an industrial vision of progress to recover their former productivity. What salmon need are humans who care, humans who pay attention to the ecosystems they live in, humans who understand the power of their stories, humans who understand that the comfort afforded by the status quo comes at a high price. Finally, the salmon need people who have been given reason to hope. Not the hope that everything will turn out right, but hope as defined by the Czech patriot Václav Havel: "the conviction that what we are doing makes sense no matter how things turn out."[33] And this can only occur when our story matches our scientific understanding of salmon-sustaining ecosystems and is consistent with how attentive and informed people experience the place where they live.

Will the citizens of Salmon Nation undertake the difficult task of writing a new salmon story, a story infused with hope, and a story that will rebuild a healthy relationship among salmon, people, and place? The answer to that question is entirely up to you.

Notes

Introduction

1 Robinson Jeffers. *Selected Poetry of Robinson Jeffers*. Stanford: Stanford University Press, 2001.

2 Callum Roberts. *The Unnatural History of the Sea*. Washington, DC: Island Press, 2007.

3 When I worked on the Olympic Peninsula (1988–1991) as a timber, fish, and wildlife biologist, most logging operations followed the rules regarding stream protection, but incidents like the one described here occurred too often.

4 Barry Lopez. *Of Wolves and Men*. New York: Charles Scribner's Sons, 1978.

5 Hans Radtke. "Hatchery Based Salmon Production in the Pacific Northwest." Portland, OR: Draft report prepared for Interrain Pacific, 1997.

6 *The Oregonian* Editorial Board. "Who's Minding Fish Store?" Portland, OR: *The Oregonian*, Section D-14, June 30, 1999.

7 John Galbraith. *The Affluent Society*. New York: New American Library, 1958.

Part I—Icebergs, Myths, and Stories

1 Mary Midgley. *The Myths We Live By*. New York: Routledge, 2004. Page 1.

2 George Lakoff. *Whose Freedom? The Battle Over America's Most Important Idea*. New York: Farrar, Straus and Giroux, 2006. Page 13.

3 John Livingston. *Arctic Oil*. Toronto: Canadian Broadcasting Corporation Merchandising, 1981.

4 One recent exception is the Independent Science Advisory Board's examination of the conceptual framework for the salmon recovery efforts in the Columbia River. See Richard Williams (editor). *Return to the River: Restoring Salmon to the Columbia River*. Burlington, MA: Elsevier Academic Press, 2006. For a discussion of the conceptual foundation and its influence on salmon management see Daniel Bottom. "To Till the Water: A History of Ideas in Fisheries Conservation." In *Pacific Salmon and Their Ecosystems: Status and Future Options*. Edited by D. Stouder, P. Bisson and R. Naiman, 569-97. New York: Chapman and Hall, 1997; Richard Williams and twelve others. "Scientific Issues in the Restoration of Salmonid Fishes in the Columbia River." *Fisheries* 24 no. 3 (1999): 10-19; and Christopher Frissell and four others. "A Resource in Crisis: Changing the Measure of Salmon Management." In *Pacific Salmon and Their Ecosystems: Status and Future Options*, Edited by D. Stouder, P. Bisson and R. Naiman, 411-44. New York: Chapman and Hall, 1997.

5 Colin Grant. *Myths We Live By*. Ottawa: University of Ottawa Press, 2001.

6 Philip Drucker. *Indians of the Northwest Coast*. Garden City, NY: The Natural History Press, 1963.

7 Grant, *Myths We Live By*; and Midgley, *The Myths We Live By*.

8 Midgley, *The Myths We Live By*. Page 2.

9   James Lichatowich and Richard Williams. "Failures to Incorporate Science into Fishery Management and Recovery Programs: Lessons from the Columbia River." In *Pacific Salmon: Ecology and Management of Western Alaska's Populations*. Edited by C. Krueger and C. Zimmerman, 1005-20. Bethesda: American Fisheries Society, Symposium 70, 2009.

10 James Lichatowich and four others. "The Existing Conceptual Foundation and the Columbia Basin Fish and Wildlife Program." In *Return to the River: Restoring Salmon to the Columbia River*. Edited by R. Williams, 29-49. Burlington, MA: Elsevier Academic Press, 2006.

11 Ibid.

12 Aldo Leopold. *A Sand County Almanac with Other Essays on Conservation from Round River*. New York: Oxford University Press, 1966.

Chapter 1—Winter Wrens and Jumbo Jets

1   Bryan Norton. *Sustainability: A Philosophy of Adaptive Ecosystem Management*. Chicago: University of Chicago Press, 2005. Page 473.

2   Edward Casey. "How to Get from Space to Place in a Fairly Short Stretch of Time." In *Sense of Place*. Edited by S. Feld and K. Basso, 13-52. Santa Fe: School of American Research Press, 1996. Pages 19 and 40.

3   Cited in David Suzuki and Wayne Grady. *Tree: A Life Story*. Vancouver, BC: Greystone Books, 2004.

4   Hilary Stewart. *Cedar: Tree of Life to the Northwest Coast Indians*. Seattle: University of Washington Press, 1984.

5   Cited in Wayne Suttles. *Coast Salish Essays*. Vancouver, BC: Talon Books, 1987. Page 26.

6   Hans Duerr. *Dreamtime: Concerning the Boundary between Wilderness and Civilization*. New York: Basil Blackwell Inc., 1978. Page 92.

7   Jim Lichatowich. *Salmon Without Rivers: A History of the Pacific Salmon Crisis*. Washington, DC: Island Press, 1999.

8   William Robbins. *Landscapes of Promise: The Oregon Story 1800-1940*. Seattle: University of Washington Press, 1997. Page 54.

9   Allerdale Grainger. *Woodsmen of the West*. Toronto: McClelland and Stewart Ltd., 1964. Page 55.

10 Herman Daly and John Cobb, Jr. *For The Common Good: Redirecting the Economy Toward Community and Environment and a Sustainable Future*. Boston: Beacon Press, 1989.

11 Lichatowich, *Salmon Without Rivers*.

12 Taken from Alfred Korzybski's statement, "The map is not the territory," quoted in Morris Berman. *The Re-Enchantment of the World*. Ithaca: Cornell University Press, 1981.

13 Gary Nabhan. *Cross Pollinations: The Marriage of Science and Poetry*. Minneapolis: Milkweed Editions, 2004.

14 John Jackson. *A Sense of Place, a Sense of Time*. New Haven: Yale University Press, 1994.

15   John Cameron. "Sense of Place, Spirit of Place: Dilemmas and Possibilities." A talk given at Barbara Blackman Temenos Seminar, Art Gallery NSW, Sydney 15/11/98. Copy available by request at www.salmonhistory.com.

16   "Landscape and Narrative" in Barry Lopez. *Crossing Open Ground.* New York: Charles Scribner's Sons, 1988.

17   Lopez, *Crossing Open Ground.*

18   Jeff Cederholm and thirteen others. "Pacific Salmon and Wildlife: Ecological Contexts, Relationships, and Implications for Management." In *Wildlife Habitat Relationships in Oregon and Washington.* D. H. Johnson and T. A. O'Neil (Manag. Dirs.). Corvallis: Oregon State University Press, 2001.

19   Casey, "How to Get From Space to Place in a Fairly Short Stretch of Time."

20   Timothy Egan. *The Good Rain: Across Time and Terrain in the Pacific Northwest.* New York: Alfred A. Knoff, 1990. Page 22.

21   Mary Willson and Karl Halupka. "Anadromous Fish as Keystone Species in Vertebrate Communities." *Conservation Biology* 9 (1995): 489-97.

22   Egan, *The Good Rain.* Page 182.

23   The Wilderness Society. *The Living Landscape Volume 2: Pacific Salmon and Federal Lands.* Washington, DC: Bolle Center for Forest Ecosystem Management, 1993.

24   John Livingston. *Rogue Primate: An Exploration of Human Domestication.* Toronto: Key Porter Books, 1994. Page 35.

25   The idea that salmon are engineered came from Rik Scarce. *Fishy Business: Salmon, Biology, and the Social Construction of Nature.* Philadelphia: Temple University Press, 2000.

26   Wes Jackson. *Becoming Native to this Place.* Washington, DC: Counterpoint Press, 1996. Page 78.

27   Eric Freyfogle. *Bounded People, Boundless Lands: Envisioning a New Land Ethic.* Washington, DC: Island Press, 1998; Michael Taylor. *Rationality and the Ideology of Disconnection.* New York: Cambridge University Press, 2006.

28   Jakob von Uexkull. "A Stroll Through the Worlds of Animals and Men: A Picture Book of Invisible Worlds." In *Instinctive Behavior,* translated and edited by C. Schiller. New York: International Universities Press, 1964.

29   David Lewis. *We the Navigators: The Ancient Art of Landfinding in the Pacific.* Honolulu: University Press of Hawaii, 1979; Stephen Thomas. *The Last Navigator.* New York: Ballantine Books, 1987.

30   Stephen Pepper. *World Hypotheses.* Berkeley: University of California Press, 1942.

31   Daniel Botkin. *Discordant Harmonies: A New Ecology for the Twenty-first Century.* New York: Oxford University Press, 1990; Max Oelschlaeger. *The Idea of Wilderness.* Binghamton: Vail-Ballou Press, 1991; and Carolyn Merchant. *The Death of Nature: Women, Ecology and Scientific Revolution.* San Francisco: Harper, 1980.

32   Cited in David Worster. *Nature's Economy: A History of Ecological Ideas.* New York: Cambridge University Press, 1985.

33   Ibid.

34   Mark Christie, Melanie Marine, Rod French, and Michael Blouin. "Genetic Adaptation to Captivity Can Occur in a Single Generation." *Proceedings of the National Academy of Science.* www.pnas.org/cgi/doi/10.1073/pnas.1111073109: 2011.

35 Samuel Hays. *Conservation and the Gospel of Efficiency: The Progressive Conservation Movement 1890-1920*. New York: Atheneum, 1969.

36 W. F. Thompson. "Fishing Treaties and Salmon of the North Pacific." *Science* 150 (1965): 1786-89.

37 Ibid.

Side Channel 1—Finding an Old Friend a Long Way from Home

1 Daniel Kozlovsky. *An Ecological and Evolutionary Ethic*. Englewood Cliffs, NJ: Prentice-Hall Inc., 1974.

Chapter 2—Salmon Stories

1 William Rees. "Net-Pen Salmon Farming: Failing on Two Fronts (And Why This is Just the Latest Stage in Humanity's Terminal Ravaging of the Seas)." In *Proceedings from World Summit on Salmon*. Editors P. Gallaher and L. Wood, 139-52. Vancouver, BC: Simon Fraser University, 2004.

2 William deBuys. *A Great Aridness: Climate Change and the Future of the American Southwest*. New York: Oxford University Press, 2011.

3 For an exhaustive treatment of societies whose stories brought crisis and disaster see Jared Diamond. *Collapse: How Societies Choose to Fail or Succeed*. New York: Viking, 2005.

4 Thomas Berry. *The Dream of the Earth*. San Francisco: Sierra Club Books, 1988.

5 In my experience, many individual Native Americans have maintained this reverence for the salmon. During the time I worked for the Jamestown S'Kallam Tribe, I had the opportunity to talk to several individual tribal members. I was impressed by those conversations. Those individuals, like their ancestors, had a strong attachment to place. They also retained many aspects of their ancestral spiritual and ethical relationships to salmon, including the sense that salmon are a gift. They retained those aspects of their culture while living in a society with very different cultural views. As one might expect, today tribal salmon management and recovery programs share many commonalities with the salmon management and recovery programs of state agencies. Approaches to tribal salmon management and recovery mimic those of Euro-Americans. Some programs are excellent, resulting in innovative ideas or producing cutting edge science while others employ the worst practices of Euro-Americans.

6 Lichatowich, *Salmon Without Rivers;* and Bottom, "To Till the Water."

7 Lynn White, Jr. "The Historical Roots of our Ecologic Crisis." *Science* 155 no. 3767 (1967): 1203-7.

8 George Goode. "The Status of the U.S. Fish Commission in 1884." Part XLI in Part XII Report of the Commission. Washington, DC: U.S. Commission of Fish and Fisheries, 1886.

9 Oregon Ecumenical Ministries. http://www.emoregon.org/environmental_ministries.php

10 Brian Czech. *Shoveling Fuel for a Runaway Train*. Berkeley: University of California Press, 2000.

11 Grant, *Myths We Live By*.

12 Robert Schoettler. "Sixty-second Annual Report of the Washington Department of Fisheries." Olympia, WA: 1953.

13 Robert Wissmar and five others. "A History of Resource Use and Disturbance in Riverine Basins of Eastern Oregon and Washington (Early 1800s-1990s)." *Northwest Science* Special Issue 68 (1994): 1-35; and B. A. McIntosh and six others."Historical Changes in Fish Habitat for Select River Basins of Eastern Oregon and Washington." *Northwest Science,* Special Issue 68 (1994): 36-53.

14 The budget is no longer reported in terms that allow identification of funds for habitat. For past spending on habitat see United States General Accounting Office (GAO). "Endangered Species: Past Actions Taken to Assist Columbia River Salmon." Washington, DC: GAO/RCED-92-173BR, 1992.

15 Cederholm, "Pacific Salmon and Wildlife."

16 In recent years some hatcheries have been placing carcasses back in the stream.

17 W. F. Thompson, "An Approach to Population Dynamics of the Pacific Red Salmon." *Transactions of the American Fisheries Society,* 88 no. 3 (1959): 206-9. Thompson used the word environment where I use habitat.

18 John Thorpe. "Performance Thresholds and Life-History Flexibility in Salmonids." *Conservation Biology.* 8, no. 3 (1994): 877-79.

19 William Liss and six others. "Developing a New Conceptual Foundation for Salmon Conservation." In *Return to the River: Restoring Salmon to the Columbia River.* Edited by R. Williams, 51-98. Burlington, MA: Elsevier Academic Press, 2006.

20 Scarce, *Fishy Business.* Page 74.

21 George Goode. "The Status of the U. S. Fish Commission in 1884."

22 James Scott. *Seeing Like a State: How Certain Schemes to Improve the Human Condition Have Failed.* New Haven, CT: Yale University Press, 1998.

23 Jim Lichatowich. "Evaluating the Performance of Salmon Management Institutions: The Importance of Performance Measures, Temporal Scales and Production Cycles." In *Pacific Salmon and Their Ecosystems: Status and Future Options.* Edited by D. Stouder, P. Bisson and R. Naiman, 69-87. New York: Chapman & Hall, 1997.

24 Robert McIntosh. *The Background of Ecology: Concept and Theory.* New York: Cambridge University Press, 1985.

25 Lichatowich, *Salmon Without Rivers.*

26 Michael Sinclair and Per Solemdal. "The Development of 'Population Thinking' in Fisheries Biology Between 1878 and 1930." *Aquatic Living Resources.* 1 no. 3 (1988): 189-213.

27 Lichatowich and Williams, "Failures to Incorporate Science into Fishery Management and Recovery Programs."

28 Ibid.

29 Lichatowich, *Salmon Without Rivers.*

30 Fred Everest. "Ecology and Management of Summer Steelhead in the Rogue River." Corvallis, OR: Oregon Game Commission, Fishery Research Report No. 7, 1973.

31 Paul Reimers. "Length of Residence of Juvenile Fall Chinook Salmon in Sixes River Oregon." Portland, OR: Research Reports of the Fish Commission of Oregon. 4 no. 2 (1973): 3-43.

32 Mark Schluchter and James Lichatowich. "Juvenile Life Histories of Rogue River Spring Chinook Salmon *Oncorhynchus tshawytscha* (Walbaum), as Determined from Scale

Analysis." Corvallis, OR: Oregon Department of Fish and Wildlife, Information Report Series, Fisheries No. 77-5, 1977.

33   By the time the biologist made his statement on the Willamette spring Chinook, other studies were also showing a diversity of life histories in Chinook salmon.

34   Joe Rojas-Burke. "Despite the Odds in the Lower Willamette, Young Salmon Grow." Portland, OR: *The Oregonian*, Section B-12, April 28, 2004.

35   Michael Reed. "A Summary of Comments Received on the Report: Biology, Behavior, and Resources of Resident and Anadromous Fish in the Lower Willamette River." Portland, OR: City of Portland Final Report of Research, 2000-2004. 2005.

36   For the 2001 effort by the legislature see Tomoko Hosaka. "GOP Leaders Want to Use Hatchery Fish to Restore Endangered Runs." Portland, OR: *The Oregonian*, Section C-1, May 23, 2001.

37   NOAA Fisheries, in response to another lawsuit, re-listed the Oregon coastal coho salmon as threatened.

38   National Oceanic and Atmospheric Administration (NOAA). "Endangered and Threatened Species: Proposed Policy on the Consideration of Hatchery-Origin Fish in Endangered Species Act Listing Determinations for Pacific Salmon and Steelhead." Washington, DC: *Federal Register* 6931354-59, 2004.

39   Ransom Myers and five others. "Hatcheries and Endangered Salmon." *Science* 303 (2004): 1980.

40   Union of Concerned Scientists. Scientific Advice on Endangered Salmon Deleted. http://www.ucsusa.org/scientific_integrity/abuses_of_science/deleting-scientific-advice-on.html

41   Willis McConnaha, Richard Williams, and James Lichatowich. "Introduction and Background of the Columbia River Salmon Problem." In *Return to the River: Restoring Salmon to the Columbia River*. Edited by R. Williams, 1-28. Burlington, MA: Elsevier Academic Press, 2006.

42   Kai Lee. *Compass and Gyroscope: Integrating Science and Politics for the Environment.* Washington, DC: Island Press, 1993.

43   Deanna Stouder, Peter Bisson and Robert Naiman (editors). *Pacific Salmon and Their Ecosystems: Status and Future Options.* New York: Chapman and Hall, 1997.

44   Marc Mangel and forty-one others. "Principles for the Conservation of Wild Living Resources." *Ecological Applications* 6 (1996): 338-72.

45   Northwest Power Planning Council. "Columbia River Basin Fish and Wildlife Program." Portland, OR, 1994.

46   Independent Scientific Review Panel (ISRP). "Review of the Columbia Basin Fish and Wildlife Program as Directed by the 1996 Amendment to the Power Act." Annual Report, ISRP 97-1. Portland, OR: Northwest Power Planning Council, 1997. Also see Lichatowich and Williams, "Failures to Incorporate Science into Fishery Management and Recovery Programs."

47   Peter Bisson and eight others. "Federal and State Approaches to Salmon Recovery at the Millennium." In *Return To the River: Restoring Salmon to the Columbia River.* Edited by R. Williams. 601-28. Burlington, MA: Elsevier Academic Press, 2006.

48   Joe Rojas-Burke. "BPA Plans to Cut Back Water Spills." Portland, OR: *The Oregonian*, Section A-1, March 31, 2004.

49 Larry Bailey and Michelle Boshard. "Follow the Money." In *Salmon 2100: The Future of Wild Pacific Salmon,* Edited by R. Lackey, D. Lach and S. Duncan, 99-124. Bethesda, MD: American Fisheries Society, 2006.

50 Eugene Hunn with James Selam and Family. *Nch'i-Wána, The Big River: Mid-Columbia Indians and Their Land.* Seattle: University of Washington Press, 1990.

51 John Allen and Marjorie Burns. *Cataclysms on the Columbia.* Portland, OR: Timber Press, 1986.

52 Larry Ward and nine others. "Elwha River Fish Restoration Plan–Developed Pursuant to the Elwha River Ecosystem and Fisheries Restoration Act, Public Law 102-495." Seattle, WA: NOAA Technical Memorandum NMFSC-90, 2008.

53 Hatchery Scientific Review Group (HSRG). "Review of the Elwha River Fish Restoration Plan and Accompanying HGMP's." A. Appleby and thirteen others. Prepared for the Lower Elwha Klallam Tribe and Washington Department of Fish and Wildlife. January 2012. Available on-line at: www.hatcheryreform.us.

54 Phillip Mundy. "Harvest Management." In *Return to the River: Restoring Salmon to the Columbia River.* Edited by R. Williams, 465-505. Burlington, MA: Elsevier Academic Press, 2006.

55 *The American Heritage Dictionary.* Third Ed. Boston: Houghton Mifflin Co., 1994.

56 Bottom, "To Till the Water."

57 Daniel Pauly and Jay Maclean. *In a Perfect Ocean: The State of Fisheries and Ecosystems in the North Atlantic Ocean.* Washington, DC: Island Press, 2003.

58 Willa Nehlsen, Jack Williams and Jim Lichatowich. "Pacific Salmon at the Crossroads: Stocks at Risk from California, Oregon, Idaho and Washington." *Fisheries* 16 no. 2 (1991): 4-21.

59 Although I have placed this statement in quotation marks, it is not a direct quote, but I have captured the exact meaning of what the individual said if not the exact words.

60 Smolts are juvenile salmon ready to migrate to sea.

61 A redd is a depression in the stream bed where the female salmon deposits her eggs.

62 National Research Council. *Upstream: Salmon and Society in the Pacific Northwest.* Washington, DC: National Academy Press, 1996.

63 Salmon Recovery Science Review Panel (SRSRP). "Report of the Meeting Held August 27-29, 2001" (Harvest Report). NOAA Fisheries, Northwest Fisheries Science Center, Seattle, WA, 2001.

64 Spencer Baird. "The Salmon Fisheries of Oregon." Portland, OR: *The Oregonian,* March 3, 1875.

65 For an excellent discussion of the importance of and problems related to escapement see Eric Knudsen. "Managing Pacific Salmon Escapements: The Gaps Between Theory and Reality." In *Sustainable Fisheries Management: Pacific Salmon.* Edited by E. Knudsen and four others, 237-72. New York: Lewis Publishers, 2000.

66 Michael Fraidenburg and Richard Lincoln. "Wild Chinook Salmon Management: An International Conservation Challenge." *North American Journal of Fisheries Management* 5 (1985): 311-29.

67 Ibid.

68 Knudsen, "Managing Pacific Salmon Escapements." Page 245.

69 Ibid. Page 247.

70 Ted Gresh, Jim Lichatowich and Peter Schoonmaker. "An Estimation of Historic and Current Levels of Salmon Production in the Northeast Pacific Ecosystem: Evidence of a Nutrient Deficit in the Freshwater Systems of the Pacific Northwest." *Fisheries* 25 no. 1 (2000): 15-21.

71 Salmon Recovery Science Review Panel, "Report for the meeting held August 27-29, 2001."

72 The exception was the harvest managers on the Skeena River. They were very cooperative.

73 This is my interpretation of the interactions with the harvest managers. Other panel members might have a different interpretation.

74 Norton, *Sustainability*.

75 Salmon Recovery Science Review Panel, "Report for the meeting held August 27-29, 2001."

76 Robert McClure, "State Salmon Harvest Gets Bad Review." Seattle, WA: *Seattle Post-Intelligencer*, December 12, 2001.

77 For a general explanation of this problem see Norton, *Sustainability*. Page 23.

78 Bottom, "To Till the Water."

79 The Independent Science Group was a panel of eleven senior scientists and managers charged with reviewing the scientific quality of the Northwest Power Planning Council's Fish and Wildlife Program.

80 Williams and twelve others, "Scientific Issues in the Restoration of Salmonid Fishes in the Columbia River."

81 Ibid.

82 Bottom, "To Till the Water."

83 G. A. Rose. "Reconciling Overfishing and Climate Change with Stock Dynamics of Atlantic Cod (*Gadus morhua*) Over 500 Years." *Canadian Journal of Fisheries and Aquatic Sciences* 61 (2004): 1553-57.

84 Alan Christopher Finlayson. *Fishing for Truth: A Sociological Analysis of Northern Cod Stock Assessment From 1977-1990*. Institute of Social and Economic Research, Study No. 52. St. Johns, NL: Memorial University of Newfoundland, 1994. Page 25.

85 Fisheries Resource Conservation Council. "Conservation Requirements of 2J3KL Cod." Report to the Minister of Fisheries and Oceans. Ottawa, On. CA: 2001.

86 Gary Meffe. "Techno-Arrogance and Halfway Technologies: Salmon Hatcheries on the Pacific Coast of North America." *Conservation Biology* 6 no. 3 (1992): 350-54.

87 Bill Monroe. "Willamette, Clackamas Salmon Runs Looking Grim." Portland: *The Sunday Oregonian*, Section C-2, December 18, 2005.

Side Channel 2—Thin Green Lines

1 Robert Ruark. *The Old Man's Boy Grows Older*. New York: Henry Holt and Company, 1957.

Chapter 3—The Meeting

1   Ulrich Beck. *World Risk Society*. Malden, MA: Blackwell Publishers Inc., 1999.

2   Theodore Roosevelt. "Eighth Annual Message." (1908) In *State of the Union Messages of the Presidents, 1790-1966*. Vol 3. Edited by Fred Israel. New York: Chelsea House Publishers, 1966.

3   Washington State Senate, Interim Investigating Committee. "Report on the Problems Affecting the Fisheries of the Columbia River." Olympia, 1943.

4   Oregon State Planning Board. "A Study of Commercial Fishing Operations on the Columbia River." Submitted to the Governor of Oregon, Salem, 1938.

5   National Research Council, *Upstream.*

6   I experienced similar attitudes when I worked for the U.S. Fish and Wildlife Service. The exception was the Jamestown S'Kallam Tribe. While I was employed by the tribe, its focus was on habitat.

7   Thompson, "An Approach to Population Dynamics of the Pacific Red Salmon."

8   G. Hartman, C. Groot and T. Northcote. "Science and Management in Sustainable Salmonid Fisheries: The Ball Is not in Our Court." In *Sustainable Fisheries Management: Pacific Salmon*. Edited by E. Knudsen, C. Steward, D. MacDonald, J. Williams and D. Reiser, 31-50. New York: Lewis Publishers, 2000.

9   Charles Wilkinson and Daniel Conner. "The Law of Pacific Salmon Fishery: Conservation and Allocation of a Transboundary Common Property Resource." *Kansas Law Review* 32 no. 1(1983): 17-109.

10 Snake River Salmon Recovery Team. "Final Recommendations to the National Marine Fisheries Service." Seattle: National Marine Fisheries Service, 1994. Page III-4. The Snake River Salmon Recovery Team was assembled by NOAA Fisheries to write the recovery plan for Snake River salmon. The team was made up of seven senior scientists including, Donald Bevan, John Harville, Peter Bergman, Theodore Bjornn, James Crutchfield, Peter Klingeman, and James Litchfield.

11 Peter Bisson, Timothy Beechie and George Pess. "Reconciling Fisheries with Conservation in Watersheds: Tools for Informed Decisions." In *Proceedings of the 4th World Fisheries Conference*. Bethesda, MD: American Fisheries Society Symposium 49 (2008): 1865-80.

12 Michael Milstein."Columbia Basin Plan Goes to Pieces." Portland, OR: *The Sunday Oregonian*, February 23, 2003.

13 B. M. Brennan. Address by B. M. Brennan, Director of Fisheries, State of Washington to the Legislative Meeting of the Columbia River Fisheries Protective Union at Astoria, Oregon. October 1, 1940. National Archives, Washington, DC: Record Group 22.

14 *The Oregonian* Editorial Board, "Who's Minding Fish Store?"

15 Scarce, *Fishy Business.*

16 Snake River Salmon Recovery Team. "Final Recommendations to the National Marine Fisheries Service."

17 Alaska Department of Fish and Game. "Pacific Salmon Rehabilitation: Highlights and Recommendations of the 1961 Governors' Conference on Salmon." Juneau, AK: Department of Fish and Game, 1961.

18 Snake River Salmon Recovery Team, "Final Recommendations to the National Marine Fisheries Service."

19   For a discussion of the changes taking place in the ocean due to climate change see J. E. N. Veron. *Reef in Time: The Great Barrier Reef from Beginning to End*. Cambridge, MA: Harvard University Press (Belknap Press), 2009.

20   The Lords of Yesterday came from Charles Wilkinson. *Crossing the Next Meridian: Land Water and the Future of the West*. Washington, DC: Island Press, 1992.

21   Blaine Harden. "Zeroing Out the Messenger." *The Washington Post,* November 30, 2005.

22   Baird, "The Salmon Fisheries of Oregon."

23   Arthur McEvoy. *The Fisherman's Problem: Ecology and Law in the California Fisheries 1850-1980*. New York: Cambridge University Press, 1986.

24   Robert Bunting. *The Pacific Raincoast: Environment and Culture in an American Eden, 1778-1900*. Lawrence: University Press of Kansas, 1997.

25   Nancy Langston. *Forest Dreams, Forest Nightmares*. Seattle: University of Washington Press, 1995. Langston wrote about forest management, but I believe the same observation is valid for fisheries management.

Side Channel 3—Visit to the River Machine

1   Donald Worster. *Rivers of Empire: Water, Aridity and the Growth of the American West*. New York: Pantheon Books, 1985.

2   Willis Rich. "Early History and Seaward Migration of Chinook Salmon in the Columbia and Sacramento Rivers." *Bulletin Bureau of Fisheries* 37 (1920): 1-74.

3   Neil Evernden. *The Natural Alien*. Toronto, ON: University of Toronto Press, 1993.

4   Jack Turner. *The Abstract Wild*. Tucson: University of Arizona Press, 1996.

Chapter 4—Coda

1   National Research Council, *Upstream*.

2   Knudsen, "Managing Pacific Salmon Escapements."

3   This discussion of the neutral tool and the idea that technologies are connected to supporting and dependent systems comes from Wolfgang Sachs. *Planet Dialectics: Exploration in Environment and Development*. Halifax, Nova Scotia: Fernwood Publishing, 1999. Quote from page 14.

4   Milo Moore, "Plan and Details for a Comprehensive Development Program of Natural Salmon Rearing Areas in the State of Washington." Prepared under contract for the Washington Department of Fisheries. Olympia, WA, 1964.

5   Livingston Stone. "Artificial Propagation of Salmon in the Columbia River Basin." *Transactions of the American Fish-Culture Association*, Washington, DC: Thirteenth Annual Meeting, May 13-14, 1884.

6   Curtis Ebbesmeyer and five others. "1976 Step in Pacific Climate: Forty Environmental Changes between 1968–1975 and 1977-1984." In *Proceedings of the Seventh Annual Pacific Climate (PACLIM) Workshop, April 1990*. Edited by J. Betancourt and V. Tharp, 115-26. California Department of Water Resources Interagency Ecological Studies Program. Technical Report 26. Sacramento, CA, 1991.

7   Williams and twelve others, "Scientific Issues in the Restoration of Salmonid Fishes in the Columbia River;" and Williams, *Return to the River*.

8   Lichatowich, *Salmon Without Rivers*.

9   Rosamond Naylor, Josh Eagle and Whitney Smith. "Salmon Aquaculture in the Pacific Northwest: A Global Industry with Local Impacts." *Environment* 45 no. 8 (2003): 18-39.

10   Steven Hume and five others. *A Stain Upon the Sea: West Coast Salmon Farming.* Madeira Park, BC: Harbour Publishing, 2004.

11   For a general discussion of communication problems between people who hold different worldviews see Bryan Norton. *Sustainability.*

12   Alexandra Morton. "Dying of Salmon Farming." In *A Stain Upon the Sea: West Coast Salmon Farming.* Steven Hume et al., 199-237. Madeira Park, BC: Harbour Publishing, 2004.

13   Ibid.

14   The problem of sea lice is described in these three papers: Morton. "Dying of Salmon Farming."; Martin Krkosek and five others. "Declining Wild Salmon Populations in Relation to Parasites from Farm Salmon." *Science,* 318(2007): 1772-75; and Martin Krkosek and four others. "Epizootics of Wild Fish Induced by Farm Fish." *Proceedings of the National Academy of Sciences* 103 no. 42 (2006): 15505-10.

15   Rachel Carson. *Silent Spring.* New York: Houghton Mifflin Company, 1962.

16   William Cronon. Foreword: "With the Best of Intentions." In *Forest Dreams, Forest Nightmares;* Williams and twelve others, "Scientific Issues in the Restoration of Salmonid Fishes in the Columbia River."; and Williams, *Return to the River.*

17   Thomas Flagg and Colin Nash (editors). "A Conceptual Framework for Conservation Hatchery Strategies for Pacific Salmonids." Seattle: National Marine Fisheries Service, NOAA Technical Memorandum NMFS-NWFSC-38, 1999.

18   Flagg and Nash, "A Conceptual Framework for Conservation Hatchery Strategies for Pacific Salmonids."; Ernie Brannon and six others. "Review of Salmonid Artificial Production of Anadromous and Resident Fish in the Columbia River Basin: Part 1 A Scientific Basis for Columbia River Production Programs." Portland, OR: Northwest Power Planning Council, Council Document 99-4, 1999. 77; also Richard Williams and three others. "Integrating Artificial Production with Salmonid Life History, Genetic, and Ecosystem Diversity: A Landscape Perspective." Portland, OR: Issue Paper for Trout Unlimited, West Coast Conservation Office, 2003; and Independent Multidisciplinary Science Team (IMST). "The Scientific Basis for Artificial Propagation in the Recovery of Wild Anadromous Salmonids in Oregon." Salem, OR: Technical Report 2001-1 to the Oregon Watershed Enhancement Board Office, 2001.

19   Regional Assessment of Supplementation Project (RASP). "Supplementation in the Columbia Basin." Portland, OR: Bonneville Power Administration. Report, Contract DE-AC06-75L01830, 1992.

20   Independent Scientific Advisory Board (ISAB). "Review of Salmon and Steelhead Supplementation." Portland, OR: Northwest Power Planning Council, ISAB 2003-3, 2003, 10.

21   Daniel Huppert. Summary (as a letter from IEAB to Chairman Cassidy) of Artificial Production Review Economics Analysis Phase 1. Portland, OR: Northwest Power and Conservation Council, Council Document IEAB 2002-1, 2002.

22   *The Oregonian* Editorial Board. "All Tangled in the Nets on the Columbia: Threatened Salmon are Dying, Sport Fishing is Crimped and an Outdoor Retail Chain Goes Belly Up." Portland, OR: *The Oregonian*, Section B-4, April 11, 2009.

23 Independent Scientific Review Panel (ISRP) and Independent Economic Analysis Board (IEAB). "SAFE Review 2007." Northwest Power and Conservation Council. Portland, OR: ISRP and IEAB Report 2007-3, 2007.

24 Donald Johnson, Wilbert Chapman and Robert Schoning."The Effects on Salmon Populations of the Partial Elimination of Fixed Fishing Gear on the Columbia River in 1935." Portland: Oregon Fish Commission Contribution no. 11. 1948.

25 Grant, *Myths We Live By.*

26 Hatchery Scientific Review Group (HSRG). Online at http://www.hatcheryreform.us/ hrp/welcome_show.action; and Artificial Production Review and Evaluation (APRE). Online at http://www.nwcouncil.org/fw/apre/Default.htm; and also, Brannon, "Review of Salmonid Artificial Production of Anadromous and Resident Fish in the Columbia River Basin."

27 Flagg and Nash, "A Conceptual Framework for Conservation Hatchery Strategies for Pacific Salmonids."; Williams and three others, "Integrating Artificial Production with Salmonid Life History, Genetic, and Ecosystem Diversity: A Landscape Perspective."; and Independent Multidisciplinary Science Team (IMST), "The Scientific Basis for Artificial Propagation in the Recovery of Wild Anadromous Salmonids in Oregon."

28 Robert Kharasch. *The Institutional Imperative: How to Understand the United States Government and Other Bulky Objects.* New York: Charterhouse Books, 1973.

29 Wilderness Society, "The Living Landscape."; and National Research Council, *Upstream.*

30 David Bella, Jonathan King and David Kailin. "The Dark Side of Organizations and a Method to Reveal It." *Emergence.* 5 no. 3 (2003): 66-82.

31 Raj Patel. *The Value of Nothing: How to Reshape Market Society and Redefine Democracy.* New York: Picador Press, 2009.

32 Bella, King and Kailin, "The Dark Side of Organization and a Method to Reveal it."

33 Text of Idaho's Department of Fish and Game code as printed in the Director's Statement on the Idaho Wildlife Summit, found online at: https://fishandgame. idaho.gov/content/article/fish-game-directors-statement-idaho-wildlife-summit.

34 Washington Department of Fish and Wildlife mission statement can be found online at: http://wdfw.wa.gov/about/mission_goals.html.

35 California mission statement can be found online at: http://www.dfg.ca.gov/about/.

Side Channel 4—A Look at the Year 2150

1 Cited in Ted Simon. *The River Stops Here: Saving Round Valley, A Pivotal Chapter in California's Water Wars.* Berkeley: University of California Press, 1994.

Part 2—Re-story-ation

1 Cited in Taylor, *Rationality and the Ideology of Disconnection.*

2 Gordon Hartman, Thomas Northcote, and Jeff Cederholm. "Human Numbers—The Alpha Factor Affecting the Future of Wild Salmon." In *Salmon 2100: The Future of Wild Pacific Salmon.* Edited by R. Lackey, D. Lach and S. Duncan, 261-92. Bethesda, MD: American Fisheries Society, 2006.

3 John Steinbeck. *Cannery Row.* New York: Viking Press, 1945.

4   Blake Gumprecht. *The Los Angeles River: Its Life, Death and Possible Rebirth*. Baltimore: Johns Hopkins University Press, 1999.

5   Ray March. *River in Ruin: The Story of the Carmel River*. Lincoln: University of Nebraska Press, 2012.

6   Wendell Berry. *Life Is a Miracle: An Essay Against Modern Superstition*. Washington, DC: Counter Point Press, 2000.

7   John Graves. *Goodbye to a River*. Austin: Texas Monthly Press, 1959. Page 58.

8   Stuart Hurlbert. "Pacific Salmon, Immigration, and Censors—Unreliability of the Cowed Technocrat." *The Social Contract Press*, 21 no. 3 (Spring 2011). http://www.thesocialcontract.com/artman2/publish/tsc_21_3/tsc-21-3-hurlbert-salmon.shtml. Also see the other papers in: *The Social Contract*, 21 no. 3 (Spring 2011) Issue Theme: "How Political Correctness Corrupts Environmental Science."

9   Hartman, Groot and Northcote, "Science and Management in Sustainable Salmonid Fisheries."

10   Hartman, Northcote and Cederholm, "Human Numbers—The Alpha Factor Affecting the Future of Wild Salmon."

11   Kenneth Ashley. "Wild Salmon in the 21st Century: Energy, Triage and Choices." In *Salmon 2100: The Future of Wild Pacific Salmon*. Edited by R. Lackey, D. Lach and S. Duncan, 71-98. Bethesda, MD: American Fisheries Society, 2006.

12   McEvoy, *The Fisherman's Problem*.

13   Lichatowich, *Salmon Without Rivers*.

14   Ibid.

15   Taken from a power point presentation prepared by Kim Hyatt and six others. "The Role of Okanagan Basin Fish and Water Management Tool in Boosting Sockeye Salmon Production." Victoria, BC: North Pacific International Chapter, American Fisheries Society Meeting, May 15, 2012.

16   Ibid.

17   Ibid.

Chapter 5—Beyond the Crossroads: First Steps Toward Salmon Recovery

1   Langston, *Forest Dreams, Forest Nightmares*.

2   Beck, *World Risk Society*.

3   Charles Warren died on March 4, 2005.

4   Williams, *Return to the River: Restoring Salmon to the Columbia River*.

5   Liss and six others, "Developing a New Conceptual Foundation for Salmon Conservation."

6   Jeffery Dose. "Commitment, Strategy, Action: The Three Pillars of Wild Salmon Recovery." In *Salmon 2100: The Future of Wild Pacific Salmon*. Edited by R. Lackey, D. Lach and S. Duncan, 233-59. Bethesda, MD: American Fisheries Society, 2006.

7   Roberts, *The Unnatural History of the Sea*.

8   For a discussion of historical myopia see Tim Smith. *Scaling Fisheries: The Science of Measuring the Effects of Fisheries 1855-1955*. New York: Cambridge University Press, 1994.

9   Board of Consultants. "Report of the Board of Consultants on the Fish Problems of the Upper Columbia River: Section 1." Palo Alto, CA: Stanford University, 1939.

10   Northwest Power Planning Council. "Artificial Production Review." Portland, OR: Northwest Power Planning Council, Report 99-15, 1999.

11   Independent Scientific Advisory Board, "Review of Salmon and Steelhead Supplementation."

12   Charles Hayes. *Beyond the American Dream: Lifelong Learning and the Search for Meaning in a Postmodern World.* Wasilla, AK: Autodidactic Press, 1998.

13   Lichatowich, "Evaluating the Performance of Salmon Management Institutions."

14   Ray Hilborn. "Can Fisheries Agencies Learn From Experience?" *Fisheries* 17 no. 4 (1992), 6-14.

15   Will Wright. *Wild Knowledge: Science, Language and Social Life in a Fragile Environment.* St. Paul: University of Minnesota Press, 1992.

16   Hilborn, "Can Fisheries Agencies Learn from Experience?"

17   David Montgomery. *King of Fish: The Thousand-Year Run of Salmon.* Boulder, CO: Westview Press, 2003.

18   Lee, *Compass and Gyroscope.* Page 148.

19   John Nagl. *Learning to Eat Soup with a Knife: Counterinsurgency Lessons from Malaya and Vietnam.* Chicago: University of Chicago Press, 2002.

20   Lee, *Compass and Gyroscope.*

21   Lichatowich and four others, "The Existing Conceptual Foundation and the Columbia Basin Fish and Wildlife Program."

22   Hilborn, "Can Fisheries Agencies Learn From Experience?"

23   Independent Scientific Review Panel (ISRP). "Retrospective Report 2007: Adaptive Management in the Columbia River." Portland, OR: Northwest Power and Conservation Council. Report ISRP2008-4, 2008.

24   Hilborn, "Can Fisheries Agencies Learn From Experience?"

25   Lichatowich and Williams, "Failures to Incorporate Science into Fishery Management and Recovery Programs."

26   Daniel Pauly. "Anecdotes and the Shifting Baseline Syndrome of Fisheries." *Trends in Ecology and Evolution,* 10 (1995): 430.

27   Bill Monroe. "Set-Aside for Chinook Fishery Should Protect Upriver Angling." Portland, OR: *The Sunday Oregonian,* Section C-8, February 14, 2010.

28   Northwest Power Planning Council. "Council Staff Compilation of Information on Salmon and Steelhead Losses in the Columbia River Basin." Portland, OR: Northwest Power Planning Council, 1986, Page 252 (see Tables 2 and 9).

29   Roberts, *The Unnatural History of the Sea.* Pages 36 and 257.

30   Independent Multidisciplinary Science Team (IMST). "Recovery of Wild Salmonids in Western Oregon Forests: Oregon Forest Practices Act Rules and the Measures in the Oregon Plan for Salmon and Watersheds." Salem, OR: Technical Report 1999-1 to the Oregon Plan for Salmon and Watersheds, Governor's Natural Resources Office, 1999. Page ii.

31   Anders Halverson. *An Entirely Synthetic Fish: How Rainbow Trout Beguiled America and Overran the World.* New Haven, CT: Yale University Press, 2010.

32   George Lakoff and Mark Johnson. *Metaphors We Live By.* Chicago: University of Chicago Press, 1980.

33 George Goode. "Pisciculture." The Encyclopaedia Britannica, XIX. 126-29. Chicago, IL: The Werner Co. 1898.

34 Williams and three others, "Integrating Artificial Production With Salmonid Life History."

35 Ibid.; and Independent Multidisciplinary Science Team (IMST), "The Scientific Basis for Artificial Propagation in the Recovery of Wild Anadromous Salmonids in Oregon."

36 Robert Bugert. "Mechanics of Supplementation in the Columbia River." *Fisheries* 25 no. 1 (1998): 11-20.

37 Independent Multidisciplinary Science Team (IMST). Letter to Kay Brown, Oregon Department of Fish and Wildlife, October 25, 2000.

38 Williams and three others, "Integrating Artificial Production With Salmonid Life History."

39 P. Paquet and fifteen others. "Hatcheries, Conservation, and Sustainable Fisheries—Achieving Multiple Goals: Results of the Hatchery Scientific Review Group's Columbia River Basin Review." *Fisheries* 36 no. 11 (2011): 547-61.

40 Keith Basso. *Wisdom Sits in Places: Landscape and Language Among the Western Apache.* Albuquerque: University of New Mexico Press, 1996.

41 Gregory Cajete. *Native Science: Natural Laws of Interdependence.* Santa Fe, NM: Clear Light Publishers, 2000.

42 Courtland Smith and Jennifer Gilden. "Assets to Move Watershed Councils from Assessment to Action." *Journal of American Water Resources Association*, 38 no. 3 (2002): 653-62.

43 Brent Steel. "Saving Wild Salmon: Moving from Symbolic Politics to Effective Policy." In *Salmon 2100: The Future of Wild Pacific Salmon.* Edited by R. Lackey, D. Lach and S. Duncan, 517-32. Bethesda, MD: American Fisheries Society, 2006.

44 Lee, *Compass and Gyroscope.*

45 Richard Williams and nine others. "Return to the River: Strategies for Salmon Restoration in the Columbia River Basin." In *Return to the River: Restoring Salmon to the Columbia River.* Edited by R. Williams, 629-66. Burlington, MA: Elsevier Academic Press, 2006.

46 John Kitzhaber. "A Plan for Breaking the Deadlock on the Columbia River." Speech given to the Portland (Oregon) City Club, October 3, 1997.

47 John Kitzhaber. Speech given to the Oregon Chapter of the American Fisheries Society, February 18, 2000.

48 Independent Multidisciplinary Science Team (IMST). "Defining and Evaluating Recovery of OCN Coho Salmon Stocks: Implications for Rebuilding Stocks Under the Oregon Plan." Corvallis, OR: Technical Report 1999-2, 1999.

49 The idea that actions can be symbolic in nature came from Steel, "Saving Wild Salmon: Moving From Symbolic Politics to Effective Policy."

50 Roberts, *The Unnatural History of the Sea.*

51 For a history of the idea of salmon refuges see Jim Lichatowich, Guido Rahr, Shuna Whidden and Cleve Steward. "Sanctuaries for Pacific Salmon." In *Sustainable Fisheries Management: Pacific Salmon.* Edited by E. Knudsen, C. Steward, D. MacDonald, J. Williams and D. Reiser, 675–86. New York: Lewis Publishers, 2000. For the most recent call for salmon refuges see Jim Lichatowich and Richard Williams. "Needed: Salmon

and Steelhead Refuges in the Pacific Northwest." *Flyfisher Magazine* (Autumn 2004): 28-31.

52 John Cobb. "Pacific Salmon Fisheries." Appendix 13 to the Report of the U. S. Commissioner of Fisheries for 1930. Washington, DC: Bureau of Fisheries Document No. 1092, 1930.

53 Scarce, *Fishy Business*.

54 Substitutability is one of three principles that underlay the belief in perpetual economic growth. The other two are efficiency and human capital. For a discussion of those factors see Czech, *Shoveling Fuel for a Runaway Train*.

55 Anyone who doubts that factory-produced salmon were meant to be a substitute for wild salmon should read Spencer Fullerton Baird's rationale for hatcheries published in 1875. See Baird, "The Salmon Fisheries of Oregon."

56 Scarce, *Fishy Business*.

57 Lichatowich, Rahr, Whidden and Steward, "Sanctuaries for Pacific Salmon."

58 For a more information, see Oregon Experience. "Oregon Wine: Grapes of Place." Oregon Public Broadcasting (OPB-TV) May 8, 2012. http://watch.opb.org/video/2232302077/

Side Channel 5—Another Look at the Year 2150

1 Alan Christopher Finlayson and Bonnie McCay. "Crossing the Threshold of Ecosystem Resilience: The Commercial Extinction of Northern Cod." In *Linking Social and Ecological Systems: Management Practices and Social Mechanisms for Building Resilience*. Edited by F. Berkes, C. Folke and J. Colding, 311–37. Cambridge: Cambridge University Press, 2000.

2 Jane Jacobs wrote books on cities, urban planning, and economics. She often challenged the conventional wisdom. Her books include: *The Death and Life of Great American Cities*. New York: Vintage Books, 1961; *Cities and the Wealth of Nations: Principles of Economic Life*. New York: Vintage Books, 1984; and *The Nature of Economies*. New York: Modern Library, 2000.

Chapter 6—Salmon, People, and Place

1 Antoine De Saint-Exupery. *The Little Prince*. New York: Harcourt Brace and Company, 1970.

2 William deBuys. *A Great Aridness*.

3 Ibid.

4 Kitzhaber, "A Plan for Breaking the Deadlock on the Columbia River."

5 Ibid.; Kitzhaber, Speech to the Oregon Chapter of the American Fisheries Society; John Kitzhaber. "A Tale of Two Rivers." Presentation to the National Conference of Trout Unlimited, August 16, 2000; John Kitzhaber. "Willamette Valley: Choices for the Future." Speech to the State of the Willamette Valley Conference. April 26, 2001; and John Kitzhaber. "Drawing a New Resource Map for a New Century." *Open Spaces* 8 no. 4 (2006).

6 Tim Bowling. *The Lost Coast: Salmon, Memory and the Death of Wild Culture*. Gibsons, BC: Nightwood Editions, 2007.

7   Norton, *Sustainability: A Philosophy of Adaptive Ecosystem Management.*

8   Ibid.

9   Jonathan Brinckman. "Salmon Tops Environmental Worries." Portland, OR: *The Sunday Oregonian.* Section A-1, December 7, 1997.

10 Taylor, *Rationality and the Ideology of Disconnection.*

11 For discussion of worldwide depletion of exploited fish populations see Ransom Meyers and Boris Worm. "Extinction, Survival or Recovery of Large Predatory Fishes." *Philosophical Transactions of the Royal Society London B. Biological Science* 360 no. 1453 (2005): 13-20; Roberts, *The Unnatural History of the Sea*; Charles Clover. *The End of the Line: How Overfishing Is Changing the World and What We Eat.* New York: The New Press, 2006; Michael Berrill. *The Plundered Seas: Can the World's Fish Be Saved?* San Francisco: Sierra Club Books, 1997; James McGoodwin. *Crisis in World Fisheries: People, Politics and Policies.* Stanford: Stanford University Press, 1990.

12 David Griffith. *The Estuary's Gift: An Atlantic Coast Cultural Biography.* University Park: Pennsylvania State University Press, 1999.

13 Kai Lee. "Sustaining Salmon: Three Principles." In *Pacific Salmon and Their Ecosystems.* Edited by D. Stouder, P. Bisson and R. Naiman, 665-75. New York: Chapman and Hall, 1997.

14 Animal Welfare Institute. Endangered Species Handbook, 2005 available at http://www.endangeredspecieshandbook.org/madagascar_biological_invertebrates.php; and see John Thompson. *The Coevolutionary Process.* Chicago: University of Chicago Press, 1994.

15 Thompson, *The Coevolutionary Process.*

16 For a discussion of the role of nutrients in salmonid ecosytems see John Stockner (editor). *Nutrients in Salmonid Ecosystems: Sustaining Production and Biodiversity.* Bethesda, MD: American Fisheries Society, Symposium 34, 2003.

17 The idea that the river becomes a great table during the salmon migration came from a Yakama Indian, Shea-wa, and was cited in Barbara Leibhardt. "Law, Environment and Social Change in the Columbia River Basin: The Yakima Indian Nation as a Case Study, 1840-1933." Berkeley: University of California at Berkeley, PhD. Dissertation, 1990.

18 The idea that stories tell us is from Brian Doyle's wonderful book *Mink River.* Corvallis: Oregon State University Press, 2010.

19 Scarce, *Fishy Business.*

20 Crawford Holling and Gary Meffe. "Command and Control and the Pathology of Natural Resource Management." *Conservation Biology* 10 no. 2 (1996): 328-37.

21 This is an example of goal displacement as discussed in Hilborn, "Can Fisheries Agencies Learn From Experience?"

22 Holling and Meffe, "Command and Control and the Pathology of Natural Resource Management."

23 Lee, "Sustaining Salmon: Three Principles."

24 Scarce, *Fishy Business.*

25 Egan, *The Good Rain.*

26 Lakoff, *Whose Freedom?*

27 Dennis Scarnecchia. "Salmon Management and the Search for Values." *Canadian Journal of Fisheries and Aquatic Sciences* 45 (1988): 2042-50.

28 See Carolyn Leopold and Luna Leopold's preface to the enlarged edition of *A Sand County Almanac*. Aldo Leopold. *A Sand County Almanac: With Essays on Conservation from Round River.* New York: Ballantine Books, 1986.

29 Leopold, *A Sand County Almanac.*

30 Thanks to Kai Lee for the example of recycling.

31 Washington Department of Fish and Wildlife and Oregon Department of Fish and Wildlife. "2012 Joint Staff Report: Stock Status and Fisheries for Fall Chinook Salmon, Coho Salmon, Chum Salmon, Summer Steelhead and White Sturgeon." Portland, OR: 2012.

32 Northwest Power Planning Council. "Council Staff Compilation of Information on Salmon and Steelhead Losses in the Columbia River Basin."

33 Havel cited in William deBuys. *The Walk.* San Antonio, TX: Trinity University Press, 2007.

Bibliography

Alaska Department of Fish and Game. "Pacific Salmon Rehabilitation: Highlights and Recommendations of the 1961 Governor's Conference on Salmon." Juneau, AK: Department of Fish and Game, 1961.

Allen, John, and Marjorie Burns. *Cataclysms on the Columbia.* Portland, OR: Timber Press, 1986.

Animal Welfare Institute. Endangered Species Handbook. 2005, available at: http://www.endangeredspecieshandbook.org/madagascar_biological_invertebrates.php.

Artificial Production Review and Evaluation (APRE). http://www.nwcouncil.org/fw/apre/Default.htm.

Ashley, Kenneth. "Wild Salmon in the 21st Century: Energy, Triage and Choices." In *The Future of Wild Pacific Salmon.* Edited by R. Lackey, D. Lach and S. Duncan, 71-98. Bethesda, MD: American Fisheries Society, 2006.

Bailey, Larry, and Michelle Boshard. "Follow the Money." In *Salmon 2100: The Future of Wild Pacific Salmon.* Edited by R. Lackey, D. Lach and S. Duncan, 99-124. Bethesda, MD: American Fisheries Society, 2006.

Baird, Spencer. "The Salmon Fisheries of Oregon." Portland, OR: *The Oregonian*, March 3, 1875.

Basso, Keith. *Wisdom Sits in Places: Landscape and Language Among the Western Apache.* Albuquerque: University of New Mexico Press, 1996.

Beck, Ulrich. *World Risk Society.* Malden, MA: Blackwell Publishers Inc., 1999.

Bella, David, Jonathan King and David Kailin. "The Dark Side of Organizations and a Method to Reveal it." *Emergence.* 5 no. 3 (2003): 66-82.

Berman, Morris. *The Re-Enchantment of the World.* Ithaca: Cornell University Press, 1981.

Berrill, Michael. *The Plundered Seas: Can the World's Fish Be Saved?* San Francisco: Sierra Club Books, 1997.

Berry, Thomas. *The Dream of the Earth.* San Francisco: Sierra Club Books, 1988.

Berry, Wendell. *Life Is a Miracle: An Essay Against Modern Superstition.* Washington, DC: Counterpoint Press, 2000.

Bisson, Peter, et al. "Federal and State Approaches to Salmon Recovery at the Millennium." In *Return to the River: Restoring Salmon to the Columbia River.* Edited by Richard Williams. 601-28. Burlington, MA: Elsevier Academic Press, 2006.

Bisson, Peter, Timothy Beechie and George Pess. "Reconciling Fisheries with Conservation in Watersheds: Tools for Informed Decisions." In *Proceedings of the 4th World Fisheries Conference.* Bethesda, MD: American Fisheries Society Symposium 49 (2008): 1865-80.

Board of Consultants. "Report of the Board of Consultants on the Fish Problems of the Upper Columbia River: Section 1." Palo Alto, CA: Stanford University, 1939.

Botkin, Daniel. *Discordant Harmonies: A New Ecology for the Twenty-first Century.* New York: Oxford University Press, 1990.

Bottom, Daniel. "To Till the Water: A History of Ideas in Fisheries Conservation." In *Pacific Salmon and Their Ecosystems: Status and Future Options.* Edited by Deanna Stouder, Peter Bisson and Robert Naiman, 569-97. New York: Chapman and Hall, 1997.

Bowling, Tim. *The Lost Coast: Salmon, Memory and the Death of Wild Culture*. Gibsons, BC: Nightwood Editions, 2007.

Brannon, Ernie, et al."Review of Salmonid Artificial Production of Anadromous and Resident Fish in the Columbia River Basin: Part 1. A Scientific Basis for Columbia River Production Programs." Portland, OR: Northwest Power Planning Council, Council Document 99-4, 1999.

Brennan, B. M. Address by B. M. Brennan, Director of Fisheries, State of Washington, to the Legislative Meeting of the Columbia River Fisheries Protective Union at Astoria, Oregon. October 1, 1940. National Archives, Washington, DC: Record Group 22.

Brinckman, Jonathan. "Salmon Tops Environmental Worries." Portland, OR: *The Sunday Oregonian*. Section A-1, December 7, 1997.

Bugert, Robert. "Mechanics of Supplementation in the Columbia River." *Fisheries* 25 no.1 (1998): 11-20.

Bunting, Robert. *The Pacific Raincoast: Environment and Culture in an American Eden, 1778-1900*. Lawrence: University Press of Kansas, 1997.

Cajete, Gregory. *Native Science: Natural Laws of Interdependence*. Santa Fe, NM: Clear Light Publishers, 2000.

Cameron, John. "Sense of Place, Spirit of Place: Dilemmas and Possibilities." A talk given at Barbara Blackman Temenos Seminar, Art Gallery NSW, Sydney, 15/11/98. Copy available by request at www.salmonhistory.com.

Carson, Rachel. *Silent Spring*. New York: Houghton Mifflin Company, 1962.

Casey, Edward. "How to Get from Space to Place in a Fairly Short Stretch of Time." In *Sense of Place*. Edited by S. Feld and K. Basso, 13-52. Santa Fe: School of American Research Press, 1996.

Cederholm, Jeff, et al. "Pacific Salmon and Wildlife: Ecological Contexts, Relationships, and Implications for Management." In *Wildlife Habitat Relationships in Oregon and Washington*. D. H. Johnson and T. A. O'Neil (Manag. Dirs.). Corvallis: Oregon State University Press, 2001.

Christie, Mark, Melanie Marine, Rod French and Michael Blouin. "Genetic Adaptation to Captivity Can Occur in a Single Generation." *Proceedings of the National Academy of Science*. www.pnas.org/cgi/doi/10.1073/pnas.1111073109: 2011.

Clover, Charles. *The End of the Line: How Overfishing Is Changing the World and What We Eat*. New York: The New Press, 2006.

Cobb, John. "Pacific Salmon Fisheries." Appendix 13 to the Report of the U. S. Commissioner of Fisheries for 1930. Washington, DC: Bureau of Fisheries Document No. 1092, 1930.

Cronon, William. Foreword: "With the Best of Intentions." In *Forest Dreams, Forest Nightmares: The Paradox of Old Growth in the Inland West*. Nancy Langston. Seattle: University of Washington Press, 1995.

Czech, Brian. *Shoveling Fuel for a Runaway Train*. Berkeley: University of California Press, 2000.

Daly, Herman, and John Cobb, Jr. *For The Common Good: Redirecting the Economy Toward Community and Environment and a Sustainable Future*. Boston: Beacon Press, 1989.

deBuys, William. *The Walk*. San Antonio, TX: Trinity University Press, 2007.

deBuys, William. *A Great Aridness: Climate Change and the Future of the American Southwest*. New York: Oxford University Press, 2011.

De Saint-Exupery, Antoine. *The Little Prince*. New York: Harcourt Brace and Company, 1970.

Diamond, Jared. *Collapse: How Societies Choose to Fail or Succeed*. New York: Viking, 2005.

Dose, Jeffery. "Commitment, Strategy, Action: The Three Pillars of Wild Salmon Recovery." In *Salmon 2100: The Future of Wild Pacific Salmon*. Edited by R. Lackey, D. Lach and S. Duncan, 233-59. Bethesda, MD: American Fisheries Society, 2006.

Doyle, Brian. *Mink River*. Corvallis: Oregon State University Press, 2010.

Drucker, Philip. *Indians of the Northwest Coast*. Garden City, NY: The Natural History Press, 1963.

Duerr, Hans. *Dreamtime: Concerning the Boundary Between Wilderness and Civilization*. New York: Basil Blackwell Inc., 1978.

Ebbesmeyer, Curtis, et al. "1976 Step in Pacific Climate: Forty Environmental Changes between 1968–1975 and 1977-1984." In *Proceedings of the Seventh Annual Pacific Climate (PACLIM) Workshop, April 1990*. Edited by J. Betancourt and V. Tharp, 115-26. California Department of Water Resources Interagency Ecological Studies Program. Technical Report 26. Sacramento: CA 1991.

Egan, Timothy. *The Good Rain: Across Time and Terrain in the Pacific Northwest*. New York: Alfred A. Knopf, 1990.

Everest, Fred. "Ecology and Management of Summer Steelhead in the Rogue River." Corvallis: Oregon Game Commission, Fishery Research Report No. 7, 1973.

Evernden, Neil. *The Natural Alien*. Toronto, ON: University of Toronto Press, 1993.

Finlayson, Alan Christopher. *Fishing for Truth: A Sociological Analysis of Northern Cod Stock Assessment from 1977-1990*. Institute of Social and Economic Research, Study No. 52. St. Johns, NL: Memorial University of Newfoundland, 1994.

Finlayson, Alan Christopher, and Bonnie McCay. "Crossing the Threshold of Ecosystem Resilience: The Commercial Extinction of Northern Cod." In *Linking Social and Ecological Systems: Management Practices and Social Mechanisms for Building Resilience*. Edited by F. Berkes, C. Folke and J. Colding, 311-37. Cambridge: Cambridge University Press, 2000.

Fisheries Resource Conservation Council. "Conservation Requirements of 2J3KC Cod." Ottawa, ON: Minister of Fisheries and Oceans, 2001.

Flagg, Thomas, and Colin Nash (editors). "A Conceptual Framework for Conservation Hatchery Strategies for Pacific Salmonids." Seattle: National Marine Fisheries Service, NOAA Technical Memorandum NMFS-NWFSC-38, 1999.

Fraidenburg, Michael, and Richard Lincoln. "Wild Chinook Salmon Management: An International Conservation Challenge." *North American Journal of Fisheries Management* 5 (1985): 311-29.

Freyfogle, Eric. *Bounded People, Boundless Lands: Envisioning a New Land Ethic*. Washington, DC: Island Press, 1998.

Frissell, Christopher, et al. "A Resource in Crisis: Changing the Measure of Salmon Management." In *Pacific Salmon and Their Ecosystems: Status and Future Options*, Edited by Deanna Stouder, Peter Bisson, and Robert Naiman, 411-44. New York: Chapman and Hall, 1997.

Galbraith, John. *The Affluent Society*. New York: New American Library, 1958.

Goode, George. "The Status of the U. S. Fish Commission in 1884." Part XLI in Part XII Report of the Commission. Washington, DC: US Commission of Fish and Fisheries, 1886.

Goode, George. "Pisciculture." *The Encyclopaedia Britannica*, XIX. 126-29. Chicago, IL: The Werner Co. 1898.

Grainger, Allerdale. *Woodsmen of the West*. Toronto: McClelland and Stewart Ltd., 1964.

Grant, Colin. *Myths We Live By*. Ottawa: University of Ottawa Press, 2001.

Graves, John. *Goodbye to a River*. Austin: Texas Monthly Press. 1959.

Gresh, Ted, Jim Lichatowich, and Peter Schoonmaker. "An Estimation of Historic and Current Levels of Salmon Production in the Northeast Pacific Ecosystem: Evidence of a Nutrient Deficit in the Freshwater Systems of the Pacific Northwest." *Fisheries* 25 no. 1 (2000): 15-21.

Griffith, David. *The Estuary's Gift: An Atlantic Coast Cultural Biography*. University Park: Pennsylvania State University Press, 1999.

Gumprecht, Blake. *The Los Angeles River: Its Life, Death and Possible Rebirth*. Baltimore, MD: Johns Hopkins University Press, 1999.

Halverson, Anders. *An Entirely Synthetic Fish: How Rainbow Trout Beguiled America and Overran the World*. New Haven, CT: Yale University Press, 2010.

Harden, Blaine. "Zeroing Out the Messenger." Washington, DC: *The Washington Post*, November 30, 2005.

Hartman, Gordon, C. Groot and Thomas Northcote. "Science and Management in Sustainable Salmonid Fisheries: The Ball Is not in Our Court." In *Sustainable Fisheries Management: Pacific Salmon*. Edited by E. Knudsen, et al. 31-50. New York: Lewis Publishers, 2000.

Hartman, Gordon, Thomas Northcote, and Jeff Cederholm. "Human Numbers—The Alpha Factor Affecting the Future of Wild Salmon." In *Salmon 2100: The Future of Wild Pacific Salmon*. Edited by R. Lackey, D. Lach and S. Duncan. 261-92. Bethesda, MD: American Fisheries Society, 2006.

Hatchery Scientific Review Group (HSRG). http://www.hatcheryreform.us/hrp/welcome_show.action.

Hatchery Scientific Review Group (HSRG). "Review of the Elwha River Fish Restoration Plan and Accompanying HGMP's." A. Appleby, et al. Prepared for the Lower Elwha Klallam Tribe and Washington Department of Fish and Wildlife. January 2012. Available on-line at: www.hatcheryreform.us

Hayes, Charles. *Beyond the American Dream: Lifelong Learning and the Search for Meaning in a Postmodern World*. Wasillia, AK: Autodidactic Press, 1998.

Hays, Samuel. *Conservation and the Gospel of Efficiency: The Progressive Conservation Movement 1890-1920*. New York: Atheneum, 1969.

Hilborn, Ray. "Can Fisheries Agencies Learn from Experience?" *Fisheries* 17 no. 4 (1992): 6-14.

Holling, Crawford, and Gary Meffe. "Command and Control and the Pathology of Natural Resource Management." *Conservation Biology* 10 no. 2 (1996): 328-37.

Hosaka, Tomoko. "GOP Leaders Want to Use Hatchery Fish to Restore Endangered Runs." Portland: *The Oregonian*, Section C-1, May 23, 2001.

Hume, Steven, et al. *A Stain Upon the Sea: West Coast Salmon Farming*. Madeira Park, BC: Harbour Publishing, 2004.

Hunn, Eugene, with James Selam and Family. *Nch'i-Wanna, The Big River: Mid-Columbia Indians and Their Land.* Seattle: University of Washington Press, 1990.

Huppert, Daniel. Summary (as a letter from IEAB to Chairman Cassidy) of Artificial Production Review Economics Analysis Phase 1. Portland, OR: Northwest Power and Conservation Council, Council Document IEAB 2002-1, 2002.

Hurlbert, Stuart. "Pacific Salmon, Immigration, and Censor—Unreliability of the Cowed Technocrat." *The Social Contract Press,* 21 no. 3 (Spring 2011). http://www.thesocialcontract.com/artman2/publish/tsc_21_3/tsc-21-3-hurlbert-salmon.shtml

Hyatt, Kim, et al. "The Role of Okanagan Basin Fish and Water Management Tool in Boosting Sockeye Salmon Production." Victoria, BC: American Fisheries Society Meeting, May 15, 2012.

Independent Multidisciplinary Science Team (IMST). "Defining and Evaluating Recovery of OCN Coho Salmon Stocks: Implications for Rebuilding Stocks Under the Oregon Plan." Corvallis, OR: Technical Report 1999-2, 1999.

Independent Multidisciplinary Science Team (IMST). "Recovery of Wild Salmonids in Western Oregon Forests: Oregon Forest Practices Act Rules and the Measures in the Oregon Plan for Salmon and Watersheds." Salem: Technical Report 1999-1 to the Oregon Plan for Salmon and Watersheds, Governor's Natural Resources Office, 1999.

Independent Multidisciplinary Science Team (IMST). Letter to Kay Brown, Oregon Department of Fish and Wildlife, October 25, 2000.

Independent Multidisciplinary Science Team (IMST). "The Scientific Basis for Artificial Propagation in the Recovery of Wild Anadromous Salmonids in Oregon." Salem: Technical Report 2001-1 to the Oregon Watershed Enhancement Board Office, 2001.

Independent Scientific Advisory Board (ISAB). "Review of Salmon and Steelhead Supplementation." Portland, OR: Northwest Power Planning Council, ISAB 2003-3, 2003.

Independent Scientific Review Panel (ISRP). "Review of the Columbia Basin Fish and Wildlife Program as Directed by the 1996 Amendment to the Power Act." Annual Report, ISRP 97-1. Portland, OR: Northwest Power Planning Council, 1997.

Independent Scientific Review Panel (ISRP). "Retrospective Report 2007: Adaptive Management in the Columbia River." Portland, OR: Northwest Power and Conservation Council. Report ISRP2008-4, 2008.

Independent Scientific Review Panel (ISRP) and Independent Economic Analysis Board (IEAB). "SAFE Review 2007." Northwest Power and Conservation Council. Portland, OR: ISRP and IEAB Report 2007-3, 2007.

Jacobs, Jane. *The Death and Life of Great American Cities.* New York: Vintage Books, 1961.

Jacobs, Jane. *Cities and the Wealth of Nations: Principles of Economic Life.* New York: Vintage Books, 1984.

Jacobs, Jane. *The Nature of Economies.* New York: Modern Library, 2000.

Jackson, John. *A Sense of Place, a Sense of Time.* New Haven: Yale University Press, 1994.

Jackson, Wes. *Becoming Native to this Place.* Washington, DC: Counterpoint Press, 1996.

Jeffers, Robinson. *Selected Poetry of Robinson Jeffers.* Stanford: Stanford University Press, 2001.

Johnson, Donald, Wilbert Chapman and Robert Schoning. "The Effects on Salmon Populations of the Partial Elimination of Fixed Fishing Gear on the Columbia River in 1935." Portland: Oregon Fish Commission Contribution no. 11. 1948.

Kharasch, Robert. *The Institutional Imperative: How to Understand the United States Government and Other Bulky Objects*. New York: Charterhouse Books, 1973.

Kitzhaber, John. "A Plan for Breaking the Deadlock on the Columbia River." Speech given to the Portland (Oregon) City Club, October 3, 1997.

Kitzhaber, John. Speech to the Oregon Chapter of the American Fisheries Society, February 18, 2000.

Kitzhaber, John. "A Tale of Two Rivers." Presentation to the National Conference of Trout Unlimited. August 16, 2000.

Kitzhaber, John. "Willamette Valley: Choices for the Future." Speech to the State of the Willamette Valley Conference. April 26, 2001.

Kitzhaber, John. "Drawing a New Resource Map for a New Century." *Open Spaces* 8 no. 4 (2006).

Knudsen, Eric. "Managing Pacific Salmon Escapements: The Gaps Between Theory and Reality." In *Sustainable Fisheries Management: Pacific Salmon*. Edited by E. Knudsen, et al. 237-72. New York: Lewis Publishers, 2000.

Kozlovsky, Daniel. *An Ecological and Evolutionary Ethic*. Englewood Cliffs: Prentice-Hall Inc., 1974.

Krkosek, Martin, et al. "Epizootics of Wild Fish Induced by Farm Fish." *Proceedings of the National Academy of Sciences* 103 no. 42 (2006): 15505-10.

Krkosek, Martin, et al. "Declining Wild Salmon Populations in Relation to Parasites from Farm Salmon." *Science,* 318 (2007): 1772-75.

Lakoff, George, and Mark Johnson. *Metaphors We Live By*. Chicago: University of Chicago Press, 1980.

Lakoff, George. *Whose Freedom? The Battle Over America's Important Idea*. New York: Farrar, Straus and Giroux, 2006.

Langston, Nancy. *Forest Dreams, Forest Nightmares*. Seattle: University of Washington Press, 1995.

Lee, Kai. *Compass and Gyroscope: Integrating Science and Politics for the Environment*. Washington, DC: Island Press, 1993.

Lee, Kai. "Sustaining Salmon: Three Principles." In *Pacific Salmon and Their Ecosystems*. Edited by D. Stouder, P. Bisson and R. Naiman, 665-75. New York: Chapman and Hall, 1997.

Leibhardt, Barbara. "Law, Environment and Social Change in the Columbia River Basin: The Yakima Indian Nation as a Case Study, 1840-1933." Berkeley: University of California at Berkeley, PhD. Dissertation, 1990.

Leopold, Aldo. *A Sand County Almanac with Other Essays on Conservation from Round River*. New York: Oxford University Press, 1966.

Leopold, Carolyn, and Luna Leopold. Preface to enlarged edition of Aldo Leopold, *A Sand County Almanac: With Essays on Conservation from Round River*. New York: Ballantine Books, 1986.

Lewis, David. *We the Navigators: The Ancient Art of Landfinding in the Pacific*. Honolulu: University Press of Hawaii, 1979.

Lichatowich, James, et al. "The Existing Conceptual Foundation and the Columbia Basin Fish and Wildlife Program." In *Return to the River: Restoring Salmon to the Columbia River*. Edited by Richard Williams, 29-49. Burlington, MA: Elsevier Academic Press, 2006.

Lichatowich, James, and Richard Williams. "Failures to Incorporate Science into Fishery Management and Recovery Programs: Lessons from the Columbia River." In *Pacific Salmon: Ecology and Management of Western Alaska's Populations*. Edited by C. Krueger and C. Zimmerman, 1005-20. Bethesda: American Fisheries Society, Symposium 70. 2009.

Lichatowich, Jim. "Evaluating the Performance of Salmon Management Institutions: The Importance of Performance Measures, Temporal Scales and Production Cycles." In *Pacific Salmon and Their Ecosystems: Status and Future Options*. Edited by Deanna Stouder, Peter Bisson and Robert Naiman. 69-87. New York: Chapman & Hall, 1997.

Lichatowich, Jim. *Salmon Without Rivers: A History of the Pacific Salmon Crisis*. Washington, DC: Island Press, 1999.

Lichatowich, Jim, and Richard Williams. "Needed: Salmon and Steelhead Refuges in the Pacific Northwest." *Flyfisher Magazine* (Autumn 2004): 28-31.

Lichatowich, Jim, Guido Rahr, Shuna Whidden and Cleve Steward. "Sanctuaries for Pacific Salmon." In *Sustainable Fisheries Management: Pacific Salmon*. Edited by E. Knudsen, et al., 675–86. New York: Lewis Publishers, 2000.

Liss, William, et al. "Developing a New Conceptual Foundation for Salmon Conservation." In *Return to the River: Restoring Salmon to the Columbia River*. Edited by Richard Williams, 51-98. Burlington, MA: Elsevier Academic Press, 2006.

Livingston, John. "Arctic Oil." Toronto, ON: Canadian Broadcasting Corporation Merchandising, 1981.

Livingston, John. *Rogue Primate: An Exploration of Human Domestication*. Toronto, ON: Key Porter Books, 1994.

Lopez, Barry. *Of Wolves and Men*. New York: Charles Scribner's Sons, 1978.

Lopez, Barry. *Crossing Open Ground*. New York: Charles Scribner's Sons, 1988.

Mangel, Marc, et al. "Principles for the Conservation of Wild Living Resources." *Ecological Applications* 6 (1996): 338-72.

March, Ray. *River in Ruin: The Story of the Carmel River*. Lincoln: University of Nebraska Press, 2012.

McClure, Robert. "State Salmon Harvest Gets Bad Review." Seattle, WA: *Seattle Post-Intelligencer*, December 12, 2001.

McConnaha, Willis "Chip," Richard Williams, and James Lichatowich. "Introduction and Background of the Columbia River Salmon Problem." In *Return to the River: Restoring Salmon to the Columbia River*. Edited by R. Williams, 1-28. Burlington, MA: Elsevier Academic Press, 2006.

McGoodwin, James. *Crisis in World Fisheries: People, Politics and Policies*. Stanford: Stanford University Press, 1990

McEvoy, Arthur. *The Fisherman's Problem: Ecology and Law in the California Fisheries 1850-1980*. New York: Cambridge University Press, 1986.

McIntosh, B. A., et al. "Historical Changes in Fish Habitat for Select River Basins of Eastern Oregon and Washington." *Northwest Science* Special Issue 68 (1994): 36-53.

McIntosh, Robert. *The Background of Ecology: Concept and Theory*. New York: Cambridge University Press, 1985.

Meffe, Gary. "Techno-Arrogance and Halfway Technologies: Salmon Hatcheries on the Pacific Coast of North America." *Conservation Biology* 6 no. 3 (1992): 350-54.

Merchant, Carolyn. *The Death of Nature: Women, Ecology and Scientific Revolution.* San Francisco: Harper, 1980.

Meyers, Ransom, and Boris Worm. "Extinction, Survival or Recovery of Large Predatory Fishes." *Philosophical Transactions of the Royal London B. Biological Science* 360 no. 1453 (2005): 13-20.

Midgley, Mary. *The Myths We Live By.* New York: Routledge, 2004.

Milstein, Michael. "Columbia Basin Plan Goes to Pieces." Portland: The *Sunday Oregonian*, February 23, 2003.

Monroe, Bill. "Set-Aside for Chinook Fishery Should Protect Upriver Angling." Portland: *The Sunday Oregonian*, Section C-8, February 14, 2010.

Monroe, Bill. "Willamette, Clackamas Salmon Runs Looking Grim." Portland: *The Sunday Oregonian*, Section C-2, December 18, 2005.

Montgomery, David. *King of Fish: The Thousand-Year Run of Salmon.* Boulder, CO: Westview Press, 2003.

Moore, Milo. "Plans and Details for a Comprehensive Development Program of Natural Salmon Rearing Areas in the State of Washington." Prepared under contract for the Washington Department of Fisheries. Olympia, Washington, 1964.

Morton, Alexandra. "Dying of Salmon Farming." In *A Stain Upon the Sea: West Coast Salmon Farming.* Steven Hume, et al, 199-237. Madeira Park, BC: Harbour Publishing, 2004.

Mottley, C. McC. Foreword in Roderick Haig-Brown. *The Western Angler.* New York: William Morrow and Company, 1947.

Mundy, Phillip. "Harvest Management." In *Return to the River: Restoring Salmon to the Columbia River.* Edited by Richard Williams, 465-505. Burlington, MA: Elsevier Academic Press, 2006.

Myers, Ransom, et al. "Hatcheries and Endangered Salmon." *Science* 303 (2004): 1980.

Nabhan, Gary. *Cross Pollinations: The Marriage of Science and Poetry.* Minneapolis, MN: Milkweed Editions, 2004.

Nagl, John. *Learning to Eat Soup with a Knife: Counterinsurgency Lessons from Malaya and Vietnam.* Chicago: University of Chicago Press, 2002.

National Oceanic and Atmospheric Administration (NOAA). "Endangered and Threatened Species: Proposed Policy on the Consideration of Hatchery-Origin Fish in Endangered Species Act Listing Determinations for Pacific Salmon and Steelhead." Washington, DC: *Federal Register* 6931354-31359, 2004.

National Research Council. *Upstream: Salmon and Society in the Pacific Northwest.* Washington, DC: National Academy Press, 1996.

Naylor, Rosamond, Josh Eagle and Whitney Smith. "Salmon Aquaculture in the Pacific Northwest: A Global Industry with Local Impacts." *Environment* 45 no. 8 (2003): 18-39.

Nehlsen, Willa, Jack Williams and Jim Lichatowich. "Pacific Salmon at the Crossroads: Stocks at Risk from California, Oregon, Idaho and Washington." *Fisheries* 16 no. 2 (1991): 4-21.

Northwest Power Planning Council. "Council Staff Compilation of Information on Salmon and Steelhead Losses in the Columbia River Basin." Portland, OR: Northwest Power Planning Council, 1986.

Northwest Power Planning Council. "Columbia River Basin Fish and Wildlife Program." Portland, OR: Northwest Power Planning Council, 1994.

Northwest Power Planning Council. "Artificial Production Review." Portland, OR, Report 99-15, 1999.

Norton, Bryan. *Sustainability: A Philosophy of Adaptive Ecosystem Management*. Chicago: University of Chicago Press, 2005.

Oelschlaeger, Max. *The Idea of Wilderness*. Binghamton: Vail-Ballou Press, 1991.

Oregon Ecumenical Ministries. http://www.emoregon.org/environmental_ministries.php

Oregon Experience. "Oregon Wine: Grapes of Place." Oregon Public Broadcasting (OPB-TV) May 8, 2012. http://watch.opb.org/video/2232302077/

Oregon State Planning Board. "A Study of Commercial Fishing Operations on the Columbia River." Submitted to the Governor of Oregon, Salem, 1938.

Oregonian Editorial Board. "All Tangled in the Nets on the Columbia: Threatened Salmon are Dying, Sport Fishing is Crimped and an Outdoor Retail Chain Goes Belly Up." Portland: *The Oregonian*, Section B-4, April 11, 2009.

Oregonian Editorial Board. "Who's Minding Fish Store?" Portland,: *The Oregonian*, Section D-14, June 30, 1999.

Paquet, P., et al. "Hatcheries, Conservation, and Sustainable Fisheries—Achieving Multiple Goals: Results of the Hatchery Scientific Review Group's Columbia River Basin Review." *Fisheries* 36 no. 11 (2011): 547-61.

Patel, Raj. *The Value of Nothing: How to Reshape Market Society and Redefine Democracy*. New York: Picador Press, 2009.

Pauly, Daniel. "Anecdotes and the Shifting Baseline Syndrome of Fisheries." *Trends in Ecology and Evolution*, 10 (1995): 430.

Pauly, Daniel, and Jay Maclean. *In a Perfect Ocean: The State of Fisheries and Ecosystems in the North Atlantic Ocean*. Washington, DC: Island Press, 2003.

Pepper, Stephen. *World Hypotheses*. Berkeley: University of California Press, 1942.

Radtke, Hans. "Hatchery Based Salmon Production in the Pacific Northwest." Portland, OR: Draft report for Interrain Pacific, 1997.

Reed, Michael. "A Summary of Comments Received on the Report: Biology, Behavior, and Resources of Resident and Anadromous Fish in the Lower Willamette River." Portland, OR: City of Portland Final Report of Research, 2000-2004. 2005.

Rees, William. "Net-Pen Salmon Farming: Failing on Two Fronts (And Why This is Just the Latest Stage in Humanity's Terminal Ravaging of the Seas)." In *Proceedings from World Summit on Salmon*. Editors P. Gallaher and L. Wood, 139-52. Vancouver, BC: Simon Fraser University, 2004.

Regional Assessment of Supplementation Project (RASP). "Supplementation in the Columbia Basin." Portland, OR: Bonneville Power Administration. Report, Contract DE-AC06-75L01830, 1992.

Reimers, Paul. "Length of Residence of Juvenile Fall Chinook Salmon in Sixes River Oregon." Portland, OR: Research Reports of the Fish Commission of Oregon. 4 no. 2 (1973): 3-43.

Rich, Willis. "Early History and Seaward Migration of Chinook Salmon in the Columbia and Sacramento Rivers." *Bulletin Bureau of Fisheries* 37 (1920): 1-74.

Robbins, William. *Landscapes of Promise: The Oregon Story 1800-1940*. Seattle: University of Washington Press, 1997.

Roberts, Callum. *The Unnatural History of the Sea*. Washington, DC: Island Press, 2007.

Rojas-Burke, Joe. "BPA Plans to Cut Back Water Spills." Section A-1, Portland: *The Oregonian*, March 31, 2004.

Rojas-Burke, Joe. "Despite the Odds in the Lower Willamette, Young Salmon Grow." Portland: *The Oregonian*, Section B-12, April 28, 2004.

Roosevelt, Theodore. "Eighth Annual Message" (1908). In *State of the Union Messages of the Presidents, 1790-1966*. Vol 3. Edited by Fred Israel. New York: Chelsea House Publishers, 1966.

Rose, G. A. "Reconciling Overfishing and Climate Change with Stock Dynamics of Atlantic Cod (*Gadus morhua*) Over 500 Years." *Canadian Journal of Fisheries and Aquatic Sciences* 61 no. 9 (2004): 1553-57.

Ruark, Robert. *The Old Man's Boy Grows Older*. New York: Henry Holt and Company, 1957.

Sachs, Wolfgang. *Planet Dialectics: Exploration in Environment and Development*. Halifax, Nova Scotia: Fernwood Publishing, 1999.

Salmon Recovery Science Review Panel (SRSRP). "Report of the Meeting Held August 27-29, 2001" (Harvest Report). NOAA Fisheries, Northwest Fisheries Science Center, Seattle, WA, 2001.

Scarce, Rik. *Fishy Business: Salmon, Biology, and the Social Construction of Nature*. Philadelphia: Temple University Press, 2000.

Scarnecchia, Dennis. "Salmon Management and the Search for Values." *Canadian Journal of Fisheries and Aquatic Sciences* 45 (1988): 2042-50.

Schoettler, Robert. "Sixty-second Annual Report of the Washington Department of Fisheries." Olympia: 1953.

Schluchter, Mark, and James Lichatowich. "Juvenile Life Histories of Rogue River Spring Chinook Salmon *Oncorhynchus tshawytscha* (Walbaum), as Determined from Scale Analysis." Corvallis, OR: Oregon Department of Fish and Wildlife, Information Report Series, Fisheries No. 77-5, 1977.

Scott, James. *Seeing Like a State: How Certain Schemes to Improve the Human Condition Have Failed*. New Haven, CT: Yale University Press, 1998.

Simon, Ted. *The River Stops Here: Saving Round Valley, A Pivotal Chapter in California's Water Wars*. Berkeley: University of California Press, 1994.

Sinclair, Michael, and Per Solemdal. "The Development of 'Population Thinking' in Fisheries Biology Between 1878 and 1930." *Aquatic Living Resources*. 1, no. 3 (1988): 189-213.

Smith, Courtland, and Jennifer Gilden. "Assets to Move Watershed Councils from Assessment to Action." *Journal of American Water Resources Association*, 38 no. 3 (2002): 653-62.

Smith, Tim. *Scaling Fisheries: The Science of Measuring the Effects of Fisheries 1855-1955*. New York: Cambridge University Press, 1994.

Snake River Salmon Recovery Team. "Final Recommendations to the National Marine Fisheries Service." Seattle: National Marine Fisheries Service, 1994.

Steel, Brent. "Saving Wild Salmon: Moving from Symbolic Politics to Effective Policy." In *Salmon 2100: The Future of Wild Pacific Salmon*. Edited by R. Lackey, D. Lach and S. Duncan, 517-32. Bethesda, MD: American Fisheries Society, 2006.

Steinbeck, John. *Cannery Row*. New York: Viking Press, 1945.

Stewart, Hilary. *Cedar, Tree of Life to the Northwest Coast Indians*. Seattle: University of Washington Press, 1984.

Stockner, John (editor). *Nutrients in Salmonid Ecosystems: Sustaining Production and Biodiversity.* Bethesda, MD: American Fisheries Society, Symposium 34, 2003.

Stone, Livingston. "Artificial Propagation of Salmon in the Columbia River Basin." *Transactions of the American Fish-Culture Association*, Washington, DC: Thirteenth Annual Meeting, May 13-14, 1884.

Stouder, Deanna, Peter Bisson and Robert Naiman (editors). *Pacific Salmon and Their Ecosystems: Status and Future Options.* New York: Chapman and Hall, 1997.

Suttles, Wayne. *Coast Salish Essays.* Vancouver, BC: Talon Books, 1987.

Suzuki, David, and Wayne Grady. *Tree: A Life Story.* Vancouver, BC: Greystone Books, 2004.

Taylor, Michael. *Rationality and the Ideology of Disconnection.* New York: Cambridge University Press, 2006.

Thomas, Stephen. *The Last Navigator.* New York: Ballantine Books, 1987.

Thompson, John. *The Coevolutionary Process.* Chicago: University of Chicago Press, 1994.

Thompson, W. F. "An Approach to Population Dynamics of the Pacific Red Salmon." *Transactions of the American Fisheries Society*, 88 no. 3 (1959): 206-9.

Thompson, W. F. "Fishing Treaties and Salmon of the North Pacific." *Science* 150 (1965): 1786-89.

Thorpe, John. "Performance Thresholds and Life-History Flexibility in Salmonids." *Conservation Biology.* 8, no. 3 (1994): 877-79.

Turner, Jack. *The Abstract Wild.* Tucson: University of Arizona Press, 1996.

Union of Concerned Scientists. Scientific Advice on Endangered Salmon Deleted. http://www.ucsusa.org/scientific_integrity/abuses_of_science/deleting-scientific-advice-on.html

United States General Accounting Office (GAO). "Endangered Species: Past Actions Taken to Assist Columbia River Salmon." Washington, DC: GAO/RCED-92-173BR, 1992.

Veron, J. E. N. *Reef in Time: The Great Barrier Reef form Beginning to End.* Cambridge, MA: Harvard University Press (Belknap Press), 2009.

von Uexkull, Jakob. "A Stroll Through the Worlds of Animals and Men: A Picture Book of Invisible Worlds." In *Instinctive Behavior.* Translated and edited by Claire Schiller. New York: International Universities Press, 1964.

Ward, Larry, et al. "Elwha River Fish Restoration Plan–Developed Pursuant to the Elwha River Ecosystem and Fisheries Restoration Act, Public Law 102-495." Seattle, WA: NOAA Technical Memorandum NMFSC-90, 2008.

Washington State Senate, Interim Investigating Committee. "Report on the Problems Affecting the Fisheries of the Columbia River." Olympia, 1943.

Washington Department of Fish and Wildlife and Oregon Department of Fish and Wildlife. "2012 Joint Staff Report: Stock Status and Fisheries for Fall Chinook Salmon, Coho Salmon, Chum Salmon, Summer Steelhead and White Sturgeon." Portland, OR: 2012.

White, Lynn Jr. "The Historical Roots of our Ecologic Crisis." *Science* 155 no. 3767 (1967): 1203-7.

Wilderness Society, The. *The Living Landscape, Volume 2: Pacific Salmon and Federal Lands.* Washington, DC: Bolle Center for Forest Ecosystem Management, 1993.

Wilkinson, Charles. *Crossing the Next Meridian: Land Water and the Future of the West.* Washington, DC: Island Press, 1992.

Wilkinson, Charles, and Daniel Conner. "The Law of the Pacific Salmon Fishery: Conservation and Allocation of a Transboundary Common Property Resource." *Kansas Law Review* 32 no. 1(1983): 17-109.

Williams, Richard (editor). *Return to the River: Restoring Salmon to the Columbia River.* Burlington, MA: Elsevier Academic Press, 2006.

Williams, Richard, et al. "Scientific Issues in the Restoration of Salmonid Fishes in the Columbia River." *Fisheries* 24 no. 3 (1999): 10-19.

Williams, Richard, James Lichatowich, Phillip Mundy and Madison Powell. "Integrating Artificial Production with Salmonid Life History, Genetic, and Ecosystem Diversity: A Landscape Perspective." Portland, OR: Issue Paper for Trout Unlimited, West Coast Conservation Office, 2003.

Williams, Richard, et al. "Return to the River: Strategies for Salmon Restoration in the Columbia River Basin." In *Return to the River: Restoring Salmon to the Columbia River.* Edited by R. Williams, 629-66. Burlington, MA: Elsevier Academic Press, 2006.

Willson, Mary, and Karl Halupka. "Anadromous Fish as Keystone Species in Vertebrate Communities." *Conservation Biology* 9 (1995): 489-97.

Wissmar, Robert, et al. "A History of Resource Use and Disturbance in Riverine Basins of Eastern Oregon and Washington (Early 1800s-1990s)." *Northwest Science Journal* Special Issue 68 (1994): 1-35.

Worster, Donald. *Nature's Economy: A History of Ecological Ideas.* New York: Cambridge University Press, 1985.

Worster, Donald. *Rivers of Empire: Water, Aridity and the Growth of the American West.* New York: Pantheon Books, 1985.

Wright, Will. *Wild Knowledge: Science, Language and Social Life in a Fragile Environment.* St. Paul, MN: University of Minnesota Press, 1992.

Index

Afognak Island, 174
Alsea Valley Alliance, 51
anchor habitats, 178
assembly line, 93, 98-99, 167
Atlantic cod, 65

Baird, Spencer, 59, 89
Basso, Keith, 165
Beck, Ulrich, 75, 145
Bella, David, 119-20
Berry, Thomas, 37, 206
Berry, Wendell, 137
biodiversity: and habitat, 42-43, 45-46, 65;
    measures to ensure, 53; as appropriate
    vocabulary, 87; and climate change,
    140; and fish factories, 160
Bottom, Dan, 63
Bowling, Tim, 112, 194
BPA (Bonneville Power Administration),
    52-54, 77
Braudel, Fernand, 47-48, 50, 53
Brennan, B. M., 81
Brice, J. J. (Commander), 174
Broughton Archipelago, 107-8
Bugert, Robert, 160
Bunting, Robert, 90
Bush, George W., 52, 54, 139

captive brood, 111, 115-16. *See also*
    supplementation
Carmel River, 135-36
Carson, Rachael, 109, 139, 205
Casey, Edward, 13
Catholic bishops, 39
cattle feedlots, 29
Cederholm, Jeff, 133
chain of habitats, 42, 79, 85, 99, 176. *See also*
    life history-habitat chain
climate change, 84, 137, 139-40
coevolved relationship, 66, 168, 198-99, 202
coevolving relationship, 201-4

Cole Rivers Hatchery, 49
Columbia Basin, 41, 53, 75, 81, 155, 172
Columbia River, 19, 39, 46, 52, 54-56, 63-64,
    76-77, 80, 82, 85, 91, 93-94, 109-14,
    126, 141-42, 146-47, 150, 156, 170, 175,
    179, 182, 189, 206: tributaries of, 94
conceptual foundation, 9-12, 19, 44, 63-65,
    104, 106, 109, 116, 146-47, 154, 204
Coos Bay, 105

Darm, Donna, 63
Darwin, Charles, 159
De Saint-Exupery, Antoine, 189
deBuys, William, 37, 189, 191
Dose, Jeffery, 148
double loop learning, 153-54, 156
Dungeness Bay, 102
Dungeness River, 152

Earth Day, 75, 109, 205
ecological consequences: of pursuit of profit,
    15; of hatcheries, 160, 164
ecological relationships, 3, 19-21, 42-43, 46,
    66-67, 89, 94, 105, 142, 167, 190
economism, 113, 195-97, 203-5
Egan, Timothy, 22-23, 203
Elk River, 161
Elwha Recovery Plan, 55
Elwha River, 55
Endangered Species Act (ESA), 9, 30, 41,
    51-52, 68, 82, 87, 104, 109, 116, 118,
    120, 135, 150, 166, 190, 192
Entiat Hatchery, 111
environmental iceberg, 9, 20, 26, 146. *See
    also* metaphors
Environmental Protection Agency, 77, 109
Euro-Americans, 15-16, 22, 26, 28, 38, 54,
    66, 73, 89, 138, 165, 174, 184, 195, 206
Evernden, Neil, 95
evolutionary history, 26, 35, 67, 98-99, 159,
    167-68, 170, 206